Governance for the Digital World

"This enlightening book provides unique insights into the governance of the digital world, and the impact of that digital world on governance of the economy and society."
—B. Guy Peters, *Maurice Falk Professor of American Government, Pittsburgh University, USA, and Former President of International Public Policy Association (IPPA)*

"Well-researched, this book is insightful and constructive. Broadly defining institutions as an ecosystem of relationships, readers gain new perspectives on hard problems. A fast, worthwhile read!"
—Vinton Cerf, *Internet pioneer, Vice President and Chief Internet Evangelist at Google*

"*Governance for the Digital World* is a highly welcome contribution to the pursuit of good governance in what till some years ago was labeled unchartered territory in the world we are living in. It is high time to get more insight into the dilemmas, intricacies, predicaments, and, last but not least, the considerable opportunities offered by digital technologies and algorithms in particular. It is of the utmost importance that for that matter this book stresses the notion of the digital commons. Irrespective of the role of state and non-state actors, eventually digital technologies pervade the daily existence of all human beings. So good governance is not a matter of choice but sheer necessity. And, as the authors show in their in-depth analysis, good governance reaches beyond the do's and don'ts of governments. It is also about the functions and interests of private corporations and small- and midsize businesses, non-governmental organizations, offline and online media, and the citizenry at large. I commend the authors for their inclusive approach to digital governance and self-governance. Their book is at the very heart of today's pivotal debate on good governance in the digital world."
—Uri Rosenthal, *Former Minister of Foreign Affairs, Former Special Envoy for Cyber Diplomacy, and Chairman Advisory Council for Science, Technology and Innovation, Netherlands*

Fernando Filgueiras • Virgílio Almeida

Governance for the Digital World

Neither More State nor More Market

Fernando Filgueiras
School of Public Policy and Government
Getulio Vargas Foundation
Brasília, Brazil

Virgílio Almeida
Berkman Klein Center
Harvard University
Cambridge, MA, USA

Department of Computer Science
Federal University of Minas Gerais
Belo Horizonte, Brazil

ISBN 978-3-030-55247-3   ISBN 978-3-030-55248-0 (eBook)
https://doi.org/10.1007/978-3-030-55248-0

© The Editor(s) (if applicable) and The Author(s), under exclusive licence to Springer Nature Switzerland AG 2021
This work is subject to copyright. All rights are solely and exclusively licensed by the Publisher, whether the whole or part of the material is concerned, specifically the rights of translation, reprinting, reuse of illustrations, recitation, broadcasting, reproduction on microfilms or in any other physical way, and transmission or information storage and retrieval, electronic adaptation, computer software, or by similar or dissimilar methodology now known or hereafter developed.
The use of general descriptive names, registered names, trademarks, service marks, etc. in this publication does not imply, even in the absence of a specific statement, that such names are exempt from the relevant protective laws and regulations and therefore free for general use.
The publisher, the authors and the editors are safe to assume that the advice and information in this book are believed to be true and accurate at the date of publication. Neither the publisher nor the authors or the editors give a warranty, expressed or implied, with respect to the material contained herein or for any errors or omissions that may have been made. The publisher remains neutral with regard to jurisdictional claims in published maps and institutional affiliations.

Cover Pattern © John Rawsterne/patternhead.com

This Palgrave Macmillan imprint is published by the registered company Springer Nature Switzerland AG.
The registered company address is: Gewerbestrasse 11, 6330 Cham, Switzerland

This book is dedicated to my sons, Miguel and Pedro, and my wife, Iara, and my stepson João Pedro, who remind me of the hope of having better days.
(Fernando)

The book is dedicated to my family and to my little grandson Martim, that helped me remember the childhood of my sons.
(Virgilio)

Preface

Goal, Theme, and Approach

The idea of writing this book originated from the observation that the digital world, driven by a collection of interrelated digital technologies, has created new relations of power, with significant impacts on society, including the economy, politics, and governments. As the digital world's development expands and accelerates, it is crucial to understand theoretical and practical aspects of digital governance, which is a multidisciplinary emergent field. It provides the way to align digital services with societal interests. The purpose of the book is to explore new frameworks, institutional arrangements, rules, and policies for governance of the digital world.

To establish a clear connection between the digital world and digital governance, it is necessary to identify and describe the main characteristics of the digital world that can be elements of governance policies and actions. The study of digital governance is concerned with a number of overarching questions. How can we design a framework of governance that enhances global systemic trust in digital technologies, while simultaneously creating opportunities for all countries to advance their strategic and economic interests in the digital world? To search for answers to this question, we examine challenges grouped into two categories: (1) how to use digital technologies to govern better, and (2) how to govern digital technologies—based on pillars such as respect for human rights, international law, and availability of meaningful opportunity—for all people and nations.

We organized the book into three main chapters. In Chap. 2, we present initial issues, definitions and structures and problems of governance for the digital world. It basically defines the book's roadmap. Chapter 3 focuses on governance issues for public and private organizations that adopt and use digital technologies in their activities. Basically, we investigate digital transformation process and algorithmic decision-making, to understand how they imply new challenges, risks, and problems for customers and citizens. Chapter 4 deals with the design of institutions needed to govern the digital world. We discuss a vision of a future where people should be kept in the governance loop and human organizations are kept as the central point of good governance. We conclude this book by summarizing the main contributions of this work.

Who Should Read This Book?

This book is intended for a broad audience concerned about the role of digital technology in society, including academics, technologists, government actors, and civil society members engaged in debates about digitization in modern life. It can also be used as part of the references for senior undergraduates and graduate courses in Political Science, Computer Science and Engineering, Law, Government, and International Relations.

Brasília, Brazil Fernando Filgueiras
Belo Horizonte, Brazil Virgílio Almeida

Acknowledgments

We are both grateful to Professor Francisco Gaetani, whose work of uniting people goes beyond his courage to solve social problems and lead organizations. We are also immensely grateful to Professor Guy Peters, who generously read a first version—still early—of this book and supported and encouraged us to submit to a strict and selective editorial process.

Fernando thanks the Federal University of Minas Gerais, where an important part of this book was built. He would also like to thank the community of National School of Public Administration (ENAP), where the idea for this book was born and was developed. Fernando would like to gratefully thank his students and colleagues of School of Public Policy and Government, Getúlio Vargas Foundation.

Virgilio would like to gratefully thank his students and colleagues at the Computer Science Department at the Federal University of Minas Gerais (UFMG) and the community of the Berkman Klein Center at Harvard University. Virgilio would like to express his gratitude to CNPq (the Brazilian Council for Scientific Research and Development) and Harvard-CAPES (Brazilian federal government agency under the Ministry of Education) Visiting Professorship, which provided partial support for his research work for this book.

Contents

1 Introduction — 1

2 The Digital World and Governance Structures — 7

3 Digital Technologies for Governance — 43

4 Governance for Digital Technologies — 75

5 Conclusions — 105

Index — 111

ABBREVIATIONS

AI	Artificial Intelligence
GCSC	Global Commission for the Stability of Cyberspace
GGE	United Nations Group of Governmental Experts on Advancing responsible State behaviour in cyberspace in the context of international security.
ICANN	Internet Corporation for Assigned Names and Numbers
ICT	Information and Communication Technologies
IEEE	Institute of Electrical and Electronic Engineers
IoT	Internet of Things
NetMundial	Global Multistakeholder Meeting on the Future of Internet Governance
OECD	Organization for Economic Co-operation and Development
UN	United Nations
UN-ITU	United Nations—International Telecommunications Union

CHAPTER 1

Introduction

Abstract This chapter gives a brief overview of the motivation for addressing the challenge of creating governance frameworks for the digital world. It also explores some of the difficulties inherent to the process of governing the digital world. From a theoretical and conceptual viewpoint, the chapter introduces actors and elements that must be part of a perspective for the governance of the digital world.

Keywords Governance • Digital world • Institutions • Platform • Algorithm • Common • Big tech

This book's purpose is to explore new frameworks, institutional arrangements, rules, and policies for governance of the digital world. As digital technology and digitization rapidly intertwine the many dimensions of society, billions of people have witnessed a quiet and seamless integration of the Internet, software, platforms, algorithms, and digital devices into their daily lives, as well as into many forms of governance and decision making in the public and private sectors. Digitization is transforming nearly every sector of the global economy and society, with immense potential benefits to healthcare, education, public safety, and more. However, conceptually and practically, these new technologies carry significant risks.

Digital technologies already guide a vast array of decisions in both the private and public sectors. For example, private technology platforms such as Amazon, Google, Facebook, Apple, Uber, Twitter, YouTube, Microsoft, and Airbnb effectively control global access to information, services, and products. They play a significant role in setting the parameters of access to markets and freedom of expression through their proprietary algorithms. The power and effect of these algorithms are not noticeable or discernible to the global populace. Public sector decisions also are increasingly dictated by algorithmic systems of governance. For instance, predictive algorithms are used by states to calculate future risk posed by inmates and have been used in sentencing decisions in court trials.

Algorithms and artificial intelligence (AI) are augmenting and replacing human decision making in immigration and refugee systems in some countries. Algorithms are used in decisions about who has access to public services and who undergoes extra scrutiny by law enforcement. Risk-assessment algorithms have been used to identify "vulnerable" children and potential victims of child sexual exploitation so that governments can act to protect them. However, the lack of adequate governance frameworks makes oversight of these processes difficult.

There is a growing anxiety and tension regarding governance of the digital world. On one hand, the digital world is viewed as an idealistic global *common* (essentially, a common or shared commodity, such as an asset or resource), with information and knowledge to be shared in spaces beyond national jurisdictions. On the other hand, the digital world is perceived as providing services and products controlled by a few technology conglomerates and a handful of countries. In between these two visions, new problems abound. Fears arise regarding monopolies that control global information, human rights violations, and the social impact of automation and AI, as well as threats to democracies, terrorism, cybercrimes, and growing extremism. Yet is there a better way to navigate these circumstances? Garrett Hardin's "The Tragedy of the Commons" offers an inspiring metaphor to explore new frameworks to govern the digital world (Hardin 1968). For instance, a better way to think of misinformation, targeted political ads, extremist videos, and biased algorithms is as polluters of common resources, such as the model proposed in the tragedy of the commons.

Governing the digital world means going beyond what legal rules impose. Digital governance is not just about establishing formal rules. It is inserted into a series of power relations, in which the different actors of the digital world relate, define policies, and build a complex framework of formal and informal norms that organize the different layers and processes involved in the construction of digital technologies and their use by society. These power relations are of interest to us in this book. The relationship between power and technology is subtle. In industrial society, the creation and control of things gave its actors influence over the political system and the economic system, making it possible to influence people's behavior. At present, we are experiencing a change in the power structure. The influence of people's behavior does not arise from the creation and control of things, but from the creation and control of information. Big techs have commodified information and created digital technologies that have a huge influence and control on people's behavior. These technology companies are nested with governments, creating a complex network of influence and power (Floridi 2015). We are in a phase when power structures are changing, and when information commodification creates new challenges for how societies can govern the digital world.

Governing the digital world means thinking about how institutions are built and how they can be effective in governing a complex world. The digital world is comprised of power relations involving governments, technological companies, civil society, and international organizations. How do these actors fit into the governance process of the digital world? How is the governance of the digital world designed? What institutions are needed in the digital world to constitute governance that results in the greatest benefit for society and that is resilient to face the different turbulent contexts that technological change carries?

The digital world is made up of various opportunities and different perspectives on how people live, communicate, think, and behave. In addition, the digital world contains various layers of communication infrastructure, software, protocols, devices, and data that constitute cyberspace, the Internet, social media platforms, applications, systems, data, and information. The digital world is a place designed by humans to connect people, information, devices, and services. The main capital of the digital world is knowledge and information. Whoever controls knowledge and information influences collective decisions and establishes new power structures. The information society requires new norms and practices to govern the digital world. Governing means to create an institutional order that promotes collective actions to steer effective results for society. This is the challenge addressed by this book: How can society create institutions that govern the digital world in a way that is beneficial to society?

As the reader will notice throughout this work, we take as a background an institutionalist perspective on the governance problem. The complex requirements for fulfilling governance require effective institutions to be created and sustained (Pierre and Peters 2005). Institutions are the set of formal and informal norms, and the shared understanding that constrains and prescribes the action of the actors (Peters 2011; Ostrom 2010; March and Olsen 1984). In this perspective, we are interested in thinking about the organizational, normative, and practical factors of governance institutions applied to the digital world. We are interested in explaining how and why governance institutions are needed and how they can be designed and sustained so that the digital world can be governed and produce a greater benefit and sustainability of digital resources for society. As we said earlier, the digital world requires governance institutions to mediate its power relations and produce societies that are more democratic, supportive, and capable of learning and managing their resources.

Governance institutions for the digital world are not simply about creating legal control or regulation. Governance for the digital world is a complex undertaking, done in non-territorial spaces with multiple actors and huge differences. Governance for the digital world means creating meaningful institutions that become collective action in the everyday life of governments, companies, and various citizens and organizations. An institution is a set of organized rules and practices, embedded in a structure of meanings and resources that are relatively invariant in the face of individual turnover and changing external circumstances. Through institutions, norms become practices and stipulate behaviors in various situations. When defining governance institutions to the digital world, we seek norms capable of establishing rules, standards, or patterns of actions according to a desired behavior of the actors in the digital arena.

For example, institutions such as the Internet Corporation for Assigned Names and Numbers (ICANN) define a protocol that makes it possible to exchange computer-to-computer information and how this information can be traced to cybersecurity. On the other hand, institutions such as the United Nations (UN) and Organization for Economic Co-operation and Development (OECD) have made efforts to establish regulations and models for the digital transformation process. The set of these institutions is very broad and diverse. Throughout this set, the objective is to define norms that are compiled and incorporated into the practice of agents, so that digital technologies can be governed.

This book explores answers—still initial and provocative—to this central question. The reflections presented in this book have a theoretical and conceptual nature, to identify the main challenges for the governance of the digital world. We do not intend to present a complete answer on how the digital world should be governed; rather, we state the challenges and central elements that must constitute a perspective for the governance of the digital world. Understanding governance for the digital world is based on the premise that governance is a political problem. In other words, we take the approach that "[u]nderstanding governance is basically a matter of understanding the nature of state–society relationships in the pursuit of collective interests" (Pierre and Peters 2005).

In the next chapter (Chap. 2), we present the initial issues, and the structures and problems of governance for the digital world. Chapter 2 is conceptual and defines the book's roadmap. In Chap. 3, we focus on governance issues for organizations that adopt and use digital technologies in their activities. Basically, we deal with how organizations—public and private—are promoting the digital transformation process and how it implies new challenges, risks, and problems for society. Note that this first approach starts from the role of digital technologies for change, in which they play a pivotal role in the new patterns of action in society. Chapter 4 deals with the design of institutions essential to governing the digital world. The design of governance institutions constitutes the creation of institutions—that is, norms and practices—that make it possible to apply the metatheoretical elements of governance to the digital world mentioned in Chap. 3. Basically, we discuss rules that are applied to the creation, use, and processes of digital technologies; the elaboration of a framework for transparency and accountability practices; and the construction of forms of coordination of the different actors in the governance process.

We conclude this book by summarizing the main contributions of this work while also questioning the role of governance as a way to preserve the future of human organization in the digital world.

References

Floridi, L. (2015). The New Grey Power. *Philosophy & Technology, 28*, 329–332. https://doi.org/10.1007/s13347-015-0206-y.

Hardin, G. (1968). The Tragedy of Commons. *Science, 162*, 1243–1248. https://doi.org/10.1126/science.162.3859.1243.

March, J., & Olsen, J. P. (1984). The New Institutionalism: Organizational Factors in Political Life. *American Political Science Review, 78*(3), 734–749. https://doi.org/10.2307/1961840.

Ostrom, E. (2010). Beyond Markets and States: Polycentric Governance of Complex Economic Systems. *American Economic Review, 100*, 1–33.

Peters, B. G. (2011). *Institutional Theory in Political Science: The New Institutionalism*. London: Continuum.

Pierre, J., & Peters, B. G. (2005). *Governing Complex Societies*. London: Palgrave Macmillan. https://doi.org/10.1057/9780230512641.

CHAPTER 2

The Digital World and Governance Structures

Abstract The utopian vision of the digital world was once viewed as a space without borders and government. As the digital world expanded and turned into a key resource for governments, businesses, individuals and organizations, it became evident the need of governance structures for the digital world. This chapter starts out the analysis of governance structures with concepts from the tragedy of the commons. It explores different elements of digital governance in order to enhance the global systemic trust in the digital world.

Keywords Governance • Cyberspace • Internet • Utopic digital world • Social media • The tragedy of the commons • Dystopia • Pollution • Algorithm • Accountability • Transparency • Coordination • Human rights

This chapter presents concepts and definitions to construct a common understanding of what the digital world entails, along with its governance challenges. The efficiency, innovation, and speed of a digitally connected world can bring benefits to almost everyone (United Nations 2019). But the digital world also brings potential downsides, such as the rise of global inequality and the social and economic exclusion of vulnerable groups of society, particularly in the developing world. In the age of digital interdependence, we need novel governance structures and mechanisms—derived

© The Author(s) 2021
F. Filgueiras, V. Almeida, *Governance for the Digital World*,
https://doi.org/10.1007/978-3-030-55248-0_2

from a multidisciplinary approach—to nurture the positive impacts of digital technologies and limit their adverse consequences.

Governance structures are political and organizational forms that outline power and govern a set of rules for managing a collective enterprise. Governance structures encompass the political aspect, including the decision-making and accountability process, and the organizational aspect, involving the bureaucratic structures of implementation. It is important to note that governance structures have this coupled aspect of politics and organizational forms (March and Olsen 1989).

Based on this premise, in this chapter we investigate: the following:

- What defines the digital world?
- What are the main problems of the digital world that demand governance structures?
- What governance and meta-theoretical elements can be applied to governing the digital world?
- What are the challenges to create a governance process to the digital world?

THE UTOPIC DIGITAL WORLD

When John Perry Barlow made "A Declaration of the Independence of Cyberspace," the perspective was that the digital world would embody a space of full freedom, of connecting people, ideas, and information (Barlow 1996). The digital world promised all freedom and a break from industrial society. As Barlow put it, "We are creating a world that all may enter without privilege or prejudice accorded by race, economic power, military force, or station of birth" (Barlow 1996). The digital world would be an anarchic space of full freedom. Digital tools would promote equality and freedom in a whole new space, disruptive with the constituted powers.

The utopian vision is that the digital world is a common world for all citizens—a space without borders and without government. Barlow's anarchist provocation suggests that the digital world is a state of nature created by humans to break all boundaries and limits. Since the Declaration, the digital world has changed radically. Technological explosion continuously transforms the world by expanding opportunities. These opportunities create a new world where technologies play a central role. Now, digital technologies change work relations, communications, social relations, the concept of property, socialization processes, and various issues of collective life. The speed with which digital technologies change society creates a context of permanent turbulence. What was once considered a utopia, a space of freedom and connection of people, has become a dystopian image and a kind of state of nature in which prevails dissent, conflict, inequality, and post-truth.

There was a reason for Declaration of the Independence of Cyberspace. The US government was creating new legislation on the regulation of communications, promising greater control of cyberspace. The Declaration, issued in 1996, was a manifesto against any form of control over the Internet. It is not possible to infer that Barlow did not foresee that the digital world would create innovative ways for business and politics, new forms of communication, and new mechanisms for doing geopolitics. A utopian vision of the digital world presupposes it as a space without hierarchies and power relations. Disruption can often mean positive changes. However, disruption can promote breakdowns in human organization, which may result in a high cost for society.

Governing the digital world means going beyond what legal rules impose. The digital world constitutes various opportunities and different perspectives on how people live, communicate, and think. In addition, the digital world encompasses the cyberspace, the Internet, digital technologies, social media platforms, and data and information. The digital world is a place designed by humans to connect people, information, devices, and services. The digital world's main capital is information and knowledge. Connecting and sharing knowledge is a key role of the digital world.

The Actual Digital World

Digital technologies, especially the Internet, algorithms, artificial intelligence, the Internet of Things (IoT), blockchain, and massive amounts of data are transforming the world, modifying how we communicate, live, and work. The digital world is driven by a collection of interrelated digital technologies that are valuable tools to create better services, and promote security, safety, and economic prosperity that benefit society as a whole. As the digital world's development expands and accelerates, it is crucial to understand theoretical and practical aspects of digital governance, which is the way to align digital services with societal interests. To establish a clear connection between the digital world and digital governance, it is necessary to identify and describe the main characteristics of the digital world that can be elements of governance policies and actions.

The starting point is the notion of cyberspace that is sometimes used interchangeably with the digital world or digital environment. Cyberspace is the collection of computing devices connected by networks, in which electronic information is stored and utilized, and communication takes place (Clark 2010). The interconnected nature of cyberspace is often associated with the Internet. The nature of cyberspace can be represented by a model with four layers:

1. people who participate in the cyber experience;
2. information that is stored, transmitted, and transformed in cyberspace;
3. logical building blocks that make up the services; and
4. physical foundations that support the logical elements.

Benkler proposes another way of viewing the digital world. The digital world, based on the Internet, can be stratified into three layers: the content, the logical, and the physical layers (Benkler 2000). At the bottom, there is the physical layer, or the infrastructure layer, that contains distributed networks that connect billions of devices. Above that is the logical infrastructure layer, which includes the management and routing functions that keep information flowing within and across networks. It also provides the unique identifiers, addresses, and protocols that ensure that the billions of devices connected to the infrastructure layer can find each other. For regulatory purposes, Kevin Werbach considers the Internet as comprised of four layers: (i) physical, (ii) logical, (iii) applications or services, and (iv) content (Werbach 2002). The two upper layers concentrate most of the society activities that take place on top of the Internet's infrastructure and logical layers.

In many references, the term cyberspace is used in military or security contexts, where it is considered the fourth domain of operation, in addition to land, sea, and aerospace. Therefore, cyberwar, information warfare, and cybersecurity are important areas and themes for governance structures. In the digital world, governance policies focus mainly on economic and social issues.

Social media is defined as web and mobile-based Internet applications that allow the creation, access, and exchange of user-generated content that is ubiquitously accessible (Kaplan and Haenlein 2010). Social media such as Facebook, Twitter, YouTube, WhatsApp, Instagram, and others have redefined global communication by offering in their platforms new tools and new media objects for users to interact in novel ways. Social media platforms have developed innovative ways to monetize data collected from user activities, with potential social consequences for their privacy. Social media platforms share a few common characteristics. The business models of social media and Internet services (such as Google, Apple, Facebook, Twitter, and Amazon) are based on extensive data collection and third-party sharing that allows them to assemble accurate user profiles (Zuboff 2019).

Social media platforms are run by proprietary algorithms that sort through large amounts of information from users and organize their life online. Examples of social media algorithms include those that manage the Facebook News Feed, the Twitter Timeline, or YouTube recommendation engine. Social media algorithms define mostly all our online daily activities as users, making decisions and promoting different behaviors. They prioritize information, classify information and actions, associate or create relationships between entities (such as friends and preferences), and filter information according to various rules or criteria (Diakopoulos 2016).

Algorithms are at the core of the digital technologies that shape society and change the way governments, businesses, and industries operate. Algorithms play an ever-growing role in the production and dissemination of news, information, and services. They also edit and shape content selection and act as gatekeepers. Criminal justice systems, banking operations, healthcare decisions, hiring and recruitment services, and college and university admissions are typical decision-making systems driven by algorithms that rank, classify, associate, and recommend services and candidates.

For example, algorithms of artificial intelligence systems such as Watson, from IBM, applied in criminal justice can predict penalties for criminals based on the analysis of the chain of actions taken by humans that led to the crime. At first glance, this is an important innovation. The criminal justice system can gain more rationality and predictability, learn from the complexity of human actions, and provide faster and more efficient justice at a lower cost. IBM Watson intends to replace human judges and reduce the possibility of bias or injustice. However, systems such as this are based on complex algorithms and proprietary systems protected by industrial secrecy. These algorithms are not auditable and incorporate new power relations based on the man-machine relationship. Digital technologies start to dictate norms and create new problems. Tests of the IBM Watson system have shown that it produces racial bias, is not accountable, and produces injustice (O'Neil 2016).

Despite this centrality in decision making, the role and impact of algorithms are largely opaque and invisible to the public. Nevertheless, the impact of automated decision making spans different sectors of society, from industry to governments. Algorithms work without much public awareness, as they increase efficiency, reduce costs, and help companies and governments offer rapid and convenient services to customers and citizens. In some cases, algorithmic decision-making systems may be detrimental to some groups of users, in particular socially vulnerable groups.

Blockchain can be described as a shared, trusted, distributed online ledger that everyone can inspect, but which no single user controls (Shackelford and Myers 2018). Blockchain technology can be used in many different applications. It enables managing digital assets in a secure and transparent way among users. For example, it can be applied to increase security and protection of IoT systems as they spread throughout government and industry. With the widespread availability of sensors and monitoring devices, human behavior is generating data and being digitized and tracked—in almost every detail. A consequence of massive data collection processes in the digital world is a phenomenon called datafication. Datafication is the transformation of daily activities of social action into online-quantified data, thus allowing for real-time monitoring, tracking, predictive analysis, and optimizing (Mayer-Schönberger and Cukier 2013). With their scale, reach, and speed, digital technologies are transforming our societies and economies.

To cope with the unprecedented challenges of the digital world, governance frameworks need to be redesigned and redefined. The actual digital world is not exactly an area of freedom. Social media and autonomous decision systems track and nudge behaviors, thereby establishing new forms of surveillance and new forms of injustices for society. There is a wealth of data and information that is collected, stored, and analyzed to create new ways of communicating and using that data to impose other forms of sociability. For example, social media disseminate hate speech, increasing the polarization of society. They do this by reinforcing behaviors and disseminating polarized and conflicting political perspectives (Sunstein 2018).

A major problem in the digital world deals with the political scenario. Technologies impact society in different ways, create new ethical problems, new challenges for organizations, and unprecedented forms of surveillance and control. Technologies create threats to democracy because of the corrosive social and political effects of disinformation that change forms of political participation (Bowers and Zittrain 2020). The Cambridge Analytica scandal cooled the world in a state of alert about the ability of social media to define election results by accessing users' private information (Isaak and Hanna 2018). This scandal added to the political agenda the need to protect privacy, and create rigid instruments for data governance and all related ethical problems. Understanding that technologies create new forms of conflict and power, governance in the digital world is a major political problem.

THE TRAGEDY OF THE DIGITAL WORLD

In the digital world, there is a complex ecosystem of knowledge, information, and algorithms, globally shared by the Internet and technology platforms. The ecosystem represents an important resource for governments, businesses, civil society, and international organizations to improve their respective organizational missions. We can look at the digital world through the lens of the Garrett Hardin's influential *Science* article, "The Tragedy of the Commons" (Hardin 1968). In the article, Hardin presented a dilemma that appears when there is unconstrained consumption of a shared resource (such as a pasture, server, or highway) by rational individuals seeking to maximize their individual gain. The consequence of this kind of behavior is the ruin or destruction of the common resource. This is the tragedy. The ruin of a common resource, due to the action of an individual or group, impacts the whole society without distinction. It is worth noting that the tragedy of the common also occurs in reverse. Instead of taking something out of the resources, it could happen when something (such as chemicals, gases, or smog) are put into the environment, as in the case of rivers and air pollution.

Artificial intelligence applications, IoT, governments as platforms, digital public services, and open government are made possible by an abundance of data, information, and services in the digital world. Some authors view the digital world as public goods. These resources and entities can be viewed as public goods, also called "the common good" that denote goods that serve all members of a community (Benkler 2017). The digital world is a place for digital commons once people, governments, and companies share the same borderless environment that lets them exchange information, access resources, and carry out applications and services (Nagle 2018).

However, we consider that the digital world is not composed of public goods, for they would demand a non-excludable condition and the availability of non-rivalrous goods. The digital world is not yet accessible to everyone, because there exist digital divides and payment terms for accessing the Internet. The digital world is composed of common-pool resources because cyberspace is non-excludable, but it can be viewed as an asset that assumes a series of rivalry conditions. One example of a rivalrous good is an online meeting with a maximum number of participants. Another example is an uninformed data collection process or an appropriation of citizens' private data. Common-pool resources exhibit characteristics of public and private goods. Common-pool resources are available to everyone for consumption and may have limited access at a high cost to society (Ostrom 1990). Common-pool resources usually are thought of in the environmental area, and the most ordinary examples are forests and irrigation systems. Common-pool resources require governance structures that are capable of creating norms that modify the behavior of individuals to manage non-excludable and rivalrous resources. The ecosystem of the Internet is a large stock of knowledge available to everyone. But it can produce rival goods, that is, its consumption can exclude other individuals from consuming or reduce the ability for other individuals to consume it. The digital world is susceptible to the tragedy of the commons, because its structure is related to non-excludable goods with characteristics of rivalrous goods. The rivalry of goods in the digital world is not intuitive. We are essentially dealing with intangible goods related to knowledge. The consumption of these resources without a governance framework can produce common consequences to all citizens. For example, the production of disinformation by groups in social media pollutes the environment, creating tensions, polarization, uncertainty and racism among other groups in society. Disinformation is profitable in social media, creating barriers to society's access to knowledge.

We can apply the problem of the tragedy of the commons to the digital world. We should analyze the current state of the digital world as a common-pool resource. The digital world is not indiscriminately accessible to everyone, despite producing rival and nonrival goods—rival and nonrival goods represent a continuum. As a common-pool resource, the digital world is based on the sharing of knowledge and information, but which may be appropriated by some organizations or governments to create applications and business models.

The overuse or inadequate appropriation of data and information resources available on the web can generate high costs for society, such as cyberattacks, disinformation problems, and difficulties for societies to face their problems. Considering the increasing digitalization of social life, the digital world becomes more susceptible to problems with common-pool resources. For example, the Under-Secretary-General of the United Nations called for a "digital ceasefire" during the SARS-COV2 pandemic, in order to preserve the actions of the World Health Organization in facing the pandemic.[1]

We assume that the digital world shares some common characteristics with the environmental world, originally the topic of the tragedy of the commons. For these common characteristics, there are digital ecosystems that can be exploited to generate all sorts of problems of collective action and impacts that affect the whole society. Information and knowledge can be viewed as *commons*, in terms of being shared resources in a complex ecosystem subjected to social dilemmas (Hess and Ostrom 2007). For example, there are behaviors and conditions in the digital world—network and server congestion, free riding, and "pollution" (such as spam and misinformation)—that are similar to dilemmas associated with other types of commons. The costs of problems of collective action—such as cyberattacks or disinformation—are paid by the whole society. In the context of this problem, the challenge is to think how the commons of the digital world should be governed, so that they benefit the whole society and reduce the costs of collective action.

We can view the digital world as a space of common-pool resources. Social media provides new forms of communication and collective action. As an IoT, digital technologies provide monitoring systems that can benefit public environments' security and safety. Artificial intelligence can promote the automation of government and firms, ensuring greater integrity and effectiveness for their services. The possibilities of the digital world to produce public benefits are abundant (Benkler 2017). However, the world of digital commons can easily be associated with the metaphor of the tragedy of the commons. Commons are shared resources that inevitably can be ruined by ungoverned exploitation. The digital world is a complex system with shared knowledge and information that are capable of influencing and changing the economy, political processes, fundamental values and characteristics of society, and forms of human organization. What defines the digital world as a space of digital commons is the way that cyberspace, the Internet, social media platforms, data, and digital

---

[1] See https://www.vox.com/world/2020/4/10/21216477/coronavirus-covid-19-pandemic-cyberattacks-digital-ceasefire-who-hack-united-nations.

technologies connect and share knowledge (Hess and Ostrom 2007). It is interesting to recall that an assumption of the *Declaration of Independence of the Cyberspace* in 1996 by John Perry Barlow was that the digital world would be a world of free knowledge without barriers or borders set by governments (Barlow 1996).

Table 2.1 shows the digital commons, their fundamental purposes, and the potential consequences of the digital world's lack of governance. A fundamental characteristic of cyberspace is its capability of connecting people, information, devices, and services. If poorly exploited and appropriated incorrectly, it can result in a set of problems, such as authoritarianism, inequalities, and various forms of injustice, ruining the possibility of generating public benefits. The problems with digital commons' use are caused by human behavior that leads to social dilemmas. The digital commons are subject to competition for use, free riding, and overharvesting. The commodification of knowledge enables misappropriation, misinformation, and other forms of problems derived from collective actions that can be configured as the tragedy of the commons. Knowledge in the digital world is non-trivial and intangible, constituting a common resource

Table 2.1 The tragedy of the digital world

Commons	Nature	Ungoverned exploitation
Cyberspace	Connect information	Information war
	Connect devices	Surveillance
	Connect people	Authoritarianism
	Connect services	Disinformation
Internet	Connect devices	Digital divide
		Digital exclusion
		Digital attacks
		Censorship
Social media platforms	Connect people	Hate speech
		Censorship
		Disinformation
		Political instability
		Divided societies
		Human right violations
Data and digital technologies	Connect services	Bias and discrimination
		Technological redlining
		Surveillance
		Privacy violation

Source: Own elaboration

available for society to promote greater solidarity and solutions to its various problems (Hess and Ostrom 2007).

If we are to generate and achieve the desired public benefits—that is, working toward and for the common good—then the digital world must be collectively governed. The hypothesis of exploitation and misuse in the digital world produces confusion, distrust, and collective losses.

The digital world presents potential for creating change and rupture. In law, digital commons create a new set of problems, where rigid legal structures must deal with complex legal problems and situations that require flexibility (Hadfield 2016). In politics, social media has promoted divided democracies and polarized political systems (Sunstein 2018). In society, the Internet has modified the mechanisms of socialization among young people, providing new spaces for interaction (Mansell 2013). In communications, the power of social media platforms has been used to produce hate speech, fake news, and limitations to freedom of speech (Mondal et al. 2017; Sunstein 2018; Gillespie 2018). In the economy, digital technologies have transformed businesses by creating a new form of capitalism based on surveillance and the ability to change consumption patterns (Zuboff 2019).

In this light, the metaphor that the tragedy of the commons offers to the digital world is useful: it provides an understanding of why we must govern a set of digital common-pool resources that are prone to exploitation. Essentially, individuals and companies—acting independently and rationally according to their own interests—carry out exploitation. These interests contradict the community's best interests, exhausting or polluting some common-pool resources (Ostrom 1990).

The possibilities opened by viewing digital commons as part of a common-pool resource can benefit the entire community in terms of expanding and diffusing connections, knowledge, services, and new forms of communication. The problem is not the exhaustion of this common, but the possibility that the results of its exploitation benefit only a few groups or destroy the community's social bonds. The common-pool resource for the digital world does not necessarily run out, but it can be appropriated too easily by individuals or groups for purely private use. For example, the danger of the concentration of power accumulated by Amazon, Google, Facebook, Twitter, Apple, and Microsoft arises from their ability to control the foundational infrastructure of our contemporary economic, informational, and political life (Rahman 2017).

The effects of digital commons on the world is to promote a sharing economy (Frenken and Schor 2017; Sutherland and Jarrahi 2018), sharing politics (Tang and Lee 2013), and sharing society (Mills and Waite 2017). The premise is that digital commons are produced by individuals, groups, and public and private organizations, and the data and information resources enable the creation of solutions for public policies, better services, and more democracy through technological innovation applied to different services (Meijer 2015; Bindu et al. 2019). Individuals, firms, and governments can use the digital commons to enhance services, create new forms of communication and business, innovate services and policies, and promote disruption and freedom.

However, the expected public benefit of the digital world depends on forms of collaboration and mechanisms of cooperation among various agents in the context of institutions. The tragedy of the commons raises the question of how to govern these common-pool resources and underscores the importance of developing intellectual tools to understand the capabilities and limits of self-governing institutions (Ostrom 1990). The tragedy of the digital world springs from the premise that the digital commons may be appropriated by some agent, promoting not the depletion of resources, but misuse due to insufficient or nonexistent institutions. The tragedy of the digital world promotes commodification, free rides, and other forms of social dilemmas (Hess and Ostrom 2007).

The tragedy of the commons produces a dystopian picture of a catastrophic world arising from the environmental crisis (Hardin 1968). The tragedy of the digital world stems from a possibility of dystopian society that misuses the digital commons to produce distorted communication, and create inequalities, new forms of domination, injustice, violence, and opacity. The tragedy of the digital world poses the question: How can we build the capacity to create institutions that govern a digital world aimed at the common good?

Governing the Digital Commons

The governing of commons has as its corollary the production of policy prescriptions that promote capacities to solve the problems of collective action (Ostrom 1990). Commons require forms of cooperation to overcome collective action dilemmas. The digital world's resources are abundant. Agents are rational and use this information to produce goods and benefits in their own interest. One of the common-pool resources available in the digital world is data, which are appropriate to produce goods. Amazon's AI capabilities use a variety of consumer information to estimate sales and influence consumer behavior. Platforms such as Facebook, Instagram, and LinkedIn use information from individuals to establish communication links and influence the behavior of these same individuals in the public sphere (Chen 2013). For example, there are political campaigns that target Facebook users with ads based on factors

> **Pollution of the Political Environment**
> An example of how the digital world is subject to the tragedy of the commons is the way misinformation and disinformation "pollute" the environment of public communication. Misinformation is false information mistakenly or inadvertently spread. Disinformation is an intentional strategy used to circulate incorrect information to create confusion. Fake news is a type of disinformation, which includes political propaganda, deep fake, and intentionally manipulated images to delegitimize mainstream media.
>
> A considerable amount of misinformation and fake news circulate through bots. They artificially construct social media trending topics, creating a network of influence and nudging citizens' behavior, promoting the delegitimization of political, social, and economic institutions.
>
> Misinformation and disinformation widen political polarization, transform the opinions and behaviors of individuals, and undermine public communication. In other words, they eradicate communication's character of creating a common understanding.
>
> Initiatives to combat fake news are carried out using digital technologies. Bots, algorithms, and artificial intelligence systems work to identify communication problems to create accountability for social media. Internet platforms usually aim at developing two classes of interventions to reduce the flow and impact of fake news: empower individuals to evaluate the fake news they encounter, and implement structural changes that prevent exposure of individuals to fake news (Lazer et al. 2018). In order to avoid breaking individuals' confidence in the media and political institutions, Internet platforms have been investing resources and technology to deal with misinformation and disinformation during political campaigns.

such as gender, location, or political allegiance. The micro-targeted advertising on social media is highly effective in persuading voters to support a given candidate (Carlos III 2018).

Solutions to the tragedy of the digital world require policy prescriptions to govern the digital commons. Solutions to the tragedy of the commons are presented in two distinct ways for creating policy prescriptions. As Ostrom points out, the first one is the "Leviathan as the only way" (Ostrom 1990). The problem of commons management will not always be solved with cooperation. One possible solution is to centralize government control over the common-pool resources. This alternative focuses on an external agency that decides agents' specific strategies. The central authority decides who can use what resources when, and how best to manage the situation.

For example, the Chinese Internet has services and applications that separate it from the rest of the world, due to the creation of its own rules, systems, and protocols that allow the central government to break with any form of control. The Chinese government establishes strict surveillance mechanisms through platforms such as WeChat. Its application in the context of the SARS-COV2 pandemic enabled the government to measure social distance by tracking personal cellphones. This possibility stems from a wide centralization and government control over the Internet (Qiang 2011).

The second alternative is "privatization as the only way." This solution means imposing private property rights when resources are held in common (Ostrom 1990). The alternative of dividing the commons is not a solution; it might not produce an acceptable outcome to the cooperation dilemma. Rational individuals will seek ways to accumulate the use of resources to increase their participation, without necessarily producing public goods.

In contrast to China, the Silicon Valley ideology in the United States believes in forms of anarchic capitalism, in which innovative companies struggle with their business models and privatize society's data and information ecosystem to build systems and applications aimed at the diffusion of services and innovations (Duff 2016). The famous motto of Mark Zuckerberg, founder of Facebook, "Move fast and break things" represents well the spirit of digital companies of the Silicon Valley. It captures the way technology businesses in the Unites States view innovation and disruption in the context of governance systems (Vardi 2018). Crises like Brexit in the United Kingdom have shown how data and information exclusively in private hands can establish other indirect forms of surveillance and directly interfere in power relations and institutional arrangements for political regimes.

The Chinese and American solutions for governance of the digital world are different and opposite, creating a new pattern of international conflict that reduces cooperation, does not solve problems with the implementation of digital technologies, and widens the scope of the conflict. The digital world today deals with an Internet governance war that affects communication and cooperation patterns (DeNardis 2014). This is reflected in a non-solution to the problems of 5G and its reach on a global scale. Government control over Internet infrastructure, due to debates on 5G, creates new international conflicts and changes the scope of privacy rules from the moment that everything connects to the Internet (DeNardis 2020).

The solution for governing the commons is not in "more State" or "more market." The process of governing the commons requires complex institutional solutions, capable of avoiding the tragedy of the commons (Ostrom 1990). Similarly, the tragedy of the digital world demands complex institutional solutions that require governance mechanisms for the production and use of digital commons. Digital governance implies creating a theory of applied human organization for subjects related to the digital world. This theoretical questioning involves searching for the regularities of the governance process among different areas and tools available in the digital world. Moreover, it involves making abstract constructs from the complexity of the digital world (Ostrom 1990).

It bears repeating that the digital governance solution is not "more State." For example, resources of the digital world in the hands of a centralized authority risk the promotion of authoritarian political regimes (Howard 2011; Bannister and Connolly 2018; Van Dijk 2012). The digital governance solution is also not "more market," because the private sector already owns the lion's share of the Internet's infrastructure and often acts as the content producer for data and information of individuals and governments. The private sector can create extremely authoritarian and exclusionary forms of citizenship (Gillespie 2018).

Digital governance solutions require robust institutional frameworks, capable of promoting formulas for allocating costs and benefits to different partners of the digital world (Gasser and Almeida 2017; Caplan and Boyd 2016; Meijer 2015). Governance mechanisms should be applied to promote resource regulation, accountable processes for society, transparency, and coordinating the various actors involved in the activities of the digital world. The avenue to solve the tragedy of the digital world is to involve multistakeholder organizational arrangements (Almeida et al. 2015) to promote solutions to the problems of collective action, even though these organizations may be subject to stress, weakness, and failures.

The challenge is to apply governance mechanisms to build solutions to the tragedy of the digital world. The overarching inquiry is how can institutions promote solidarity, management, and capacity so that they are accountable to citizens, and adapt and learn from their own mistakes and promote values that spread development? Before delving into the merits of what is digital governance, we first must ask: What is governance?

What Is Governance?

Synthetically, governance is an institutional solution to collective action problems. Governance is an arrangement of polycentric human organization in the context of institutional grammars (Ostrom 2010; Crawford and Ostrom 1995). The digital world demands that digital commons circulating in cyberspace be governed to produce public benefits. The solution to the collective problems of the digital world will not stem from the greater centralization and control of the State and from complete privatization.

The digital world demands governance mechanisms that foster cooperation among stakeholders and build a democratic and inclusive perspective that enables the digital commons to be used in a way that creates solidarity, social justice, and transparency and accountability to society. Governance of the digital world challenges governments and societies and demands policies that steer the use of information and communication technologies (ICT) to produce public benefits. In the framework of a complex and hyperconnected digital world, new governance models need to be developed to offer viable institutional solutions.

Governance became a theoretical paradigm for governments. This paradigm is identified with a new set of practices for the governing (Stoker 2018; Peters 2010). The first element of this paradigm of governance is the steering dynamics of governing. The government does not come to be exercised in a centralized and hierarchical structure, but with the involvement of local, regional, and transnational actors (Kennett 2010). Governance should be a polycentric institutional arrangement that governs public issues (Ostrom 2010).

The actors in a polycentric arrangement are diverse organizations from the public and private sectors. They comprise social actors organized into policy networks (Rhodes 1997, 2007), bureaucratic actors (Peters and Pierre 2016), and international organizations (Stone 2008; Rosenau 1995). What delimits the governance paradigm is the introduction of organizational changes in public administration and companies. For example, decision-making systems must involve multiple stakeholders, forming networks between public, private sectors, and non-governmental organizations (NGOs).

The governance paradigm recognizes that institutions need adaptive complexity to address the various new challenges and to provide the ingredients for emerging solutions. Governance assumes that organizations should exploit interactive capacity to provide a stronger indication that governance challenges are likely to be met (Stoker 2019). Governance assumes the interactive capacity to deal with the different stakeholders and involve them in a perspective of cooperation with different parts of society.

New governance paradigms, therefore, should imply changes in the practice of governing, considering the interdependence of the various actors and constructing new identities that express solidarity. Governance also implies the expectation that citizens can participate more in collective decisions and the role of government in promoting and steering governance processes that have the capacity and legitimacy to produce rules, establish shared understandings, and coordinate this structure for the production of outcomes (March and Olsen 1989; Peters and Pierre 2016; Koppenjan and Klijn 2004; Sörensen and Torfing 2005; Stoker 1997; Salamon 2002).

Working from this viewpoint, governance reforms have been promoted by international organizations since the 1990s. These reforms are advised by organizations such as OECD and development banks and taken by several countries to increase the coherence and efficiency of public policies and to create procedures that reinforce legitimacy and accountability (World Bank 1997; OECD 2001). These reform principles also apply to corporate governance, including private sector organizations to increase transparency and accountability to society (OECD 2014). Today, governance reforms represent a set of complex institutional changes, in order to increase the efficiency and effectiveness of public policies, as well as more robust democratic procedures for governing.

In other words, the governance paradigm requires, on the one hand, reflecting on the political aspects that make legitimate decisions possible for the community and thereby creating shared understandings about the various social problems and mediating interests. On the other hand, the governance paradigm requires management instruments that enable the correct implementation and evaluation of public policies (Peters and Pierre 2016).

The solution for collective problems and public policies is the purpose of governance in the institutional context. The governance paradigm is based on a pluralist state (Powell 1990; Powell and DiMaggio 1991). The complex, plural, and fragmented nature of public service delivery and public policy implementation requires a governance structure based on permanent negotiation of values, meanings, and relationships, in view of multiple organizations within the context of political environments. The objective of governance is to produce institutional solutions to public problems and achieve a model of human organization that promotes solidarity, trust, and cooperation between multiple agents of government and the private sector. In this paradigm of governing, public and private institutions work together not only in the decision-making processes that affect society, but also in the implementation process.

New governance models have introduced changes in organizations. Transformations were introduced to ensure greater autonomy of organizations, which is associated with greater control over governments. Autonomy does not mean a technocracy that governs, but a decentered state that encompasses the political role of leaders, multiple stakeholders, and management elements in a nonhierarchical bureaucracy (Peters 2004; Peters and Pierre 2000). To promote a broader understanding of governance changes, scholars have promoted the prospect of meta-governance. The goal is to focus on the process of institutional changes by looking at their constituent elements, namely: hierarchies, markets, and networks (Mueleman 2008), governance instruments (Peters 2010), governance failures (Jessop 2002; Bell and Park 2006), political controls, and decision-making processes at the societal level (Sörensen and Torfing 2005).

Meta-governance studies indicate three essential and common elements of governance structures.

- *Decision making*. Institutions associated with the conventional way of governing present problems to make decisions. Representative democracy presents failures for service delivery and public policies and shows difficulties in producing decisions. Governance changes introduce new forms of political participation that promote the organization of society and deliberative capacities that can influence decision processes. Governance requires participation and deliberation mechanisms that align the public interest with government planning. Political participation and deliberation mechanisms enhance governance legitimacy and serve to create a perception of greater equity in decision-making processes (Fung 2015). Participation expands democratic inclusion and generates community perception of greater fairness and justice. From a governance perspective, public administration is a place of democratization of the State (Warren 2009).

- *Accountability and transparency.* Transparency challenges the government to provide accurate information to citizens (Heald 2006). Transparency is linked to accountability, as both require political responsibility. Political responsibility is the government action's adherence to public interest (Filgueiras 2016). Political responsibility is impossible without transparent institutions that allow a minimum institutional information deficit to citizens. Governance presupposes free knowledge from the citizen. In this case, for this knowledge to be possible, it is crucial that public organizations are clear, informative, and accountable to the citizen. Transparency becomes an instrumental value toward the exercise of accountability (Heald 2006). To preserve the public interest, accountability institutions must be autonomous and recognized by their personnel as bearers of public authority for the supervision, control, correction, and punishment of illegal acts (Olsen 2017). Transparency and accountability are key to strengthening public integrity and corruption control, and promote compliance with institutional rules.
- *Coordination.* The solution to the constant problems identified by different stakeholders in relation to public policies is related not only to the availability of information, but also to coordination during formulation and implementation of public policies. This problem is not restricted to governments; it appears in all large organizations, whether public or private (Peters 1998). Citizens and the business community need to go from public agency to agency to have access to the benefits to which they are entitled. Many government programs are contradictory, and others may have gaps that fail to provide public services. These failures result from coordination difficulties (Peters 2004; Bouckaert et al. 2010). Going further, governance reforms have a mantra: governments must improve coordination. The theoretical postulate is that coordination improves government's effectiveness.

These three elements are common to governance prescriptions, which are combined with management tools steered to performance, strategic budget, and personnel management, and soft law valuing informal interactions between government and society (Peters 2010). The concept of meta-governance makes it possible to identify the central elements to make institutional changes, considering the multiplicity and differentiation of experiences. The goal of governance is to promote the involvement of a judicious mix of market, hierarchy, and networks to achieve the best possible outcomes in services and policy to create public good and values (Stoker 1997).

The concept of governance is a new way of governing that is not state-centric or market-centric. It is polycentric and extends to the public sector, private sector, and nonprofit organizations. In this way, the concept of governance has been added by different adjectives, which qualify the sphere of action and specificity of a given sector (Ostrom 2010). Corporate governance applies primarily to firms. It is the system of rules, practices, and processes that organize the firm's direction (Williamson 2002). Corporate governance relates to the diverse—and sometimes hidden—investors and shareholders, who act as multiple principals. Corporate governance takes decisions that affect these multiple principals and require controls and risk management over decisions made (Pagano and Volpin 2005). Public governance is the new form of governing applied in public administration. It is a pluralist conception of the state in which public organizations are focused on their institutional context, and which emphasizes the process of negotiating values, meanings, and relationships in the process of building public policies and public services (Osborne 2010). The public values are plural and contested and the services are carried out in dynamic co-production and co-creation (Osborne et al. 2018; Bovaird et al. 2015). Global governance refers to the process of building cooperation among the various transnational actors (Kennett 2010). Globalization produces new forms of relationships and interaction between the State and society, providing the complexity and interdependence between international and local actors (Rosenau 1999). Collective problems are not only local, but extend to the global scenario, demanding solutions that seek cooperation among actors who are interdependent (Nye and Donahue 2000).

Governance solutions are key to the tragedy of the digital world. The key is to provide solutions to problems derived from dilemmas of collective actions in the digital world. Actors in the digital world are interdependent, rational, and seek to maximize the benefits of appropriating the digital commons. Governance must devise solutions that make it possible for the various actors to cooperate and share resources among all (Ostrom 1990). The possibility of public benefit in the digital world will occur when resources are shared, so that interdependent actors can participate in decisions and become partners, act transparently, and promote compliance with institutional rules.

What Is Digital Governance?

The digital world presents several challenges to governance structures. The tragedy of the digital world demands structures of governance that can establish human organization to promote cooperation, trust, and solidarity. The instruments of digital governance are based on the use and consequences of digital technologies for society, politics, and the economy. However, human organization, in its broad sense, depends on building institutions for the digital world, focusing on constituting rules and policies that ensure outcomes for the organization of society, the State, and the economy. In addition, structures of governance aim to establish the entire system in which public decisions are made and enacted, including institutions, norms, laws, and policies for use by the connected world's diverse resources—such as data, information, and services—in the public interest.

The issue of digital governance is new and there is still no accurate perspective on the concept. There are many definitions of digital governance, without a consensus in the academic literature. As proposed by Floridi, "Digital governance is the practice of establishing and implementing policies, procedures, and standards for the proper development, use, and management of the infosphere" (Floridi 2018). In the context of government, digital governance also can be viewed as the technology that mediates services that facilitate a transformation in the relationship between government and citizen (Oakley 2010). E-governance can be an internally focused use of ICT to manage organizational resources and administer policies and procedures (Palvia and Sharma 2007). It also means the use of ICT to improve the quality of services and governance (Chen and Hsieh 2009). Digital governance also can be defined as the application of ICT tools in (1) the interaction between government, citizens, and businesses, and (2) in internal government operations to simplify and improve democratic governance (UNPAN 2011). Finally, digital governance is an institutional framework for managing digital transformation in organizations and producing new forms of interaction between citizens and states (Welchman 2015).

None of these definitions completely fits the digital era's needs. The tragedy of the digital world presents challenges that demand a new approach for designing the concept of digital governance. The following characteristics represent challenges for continuously constructing the global digital governance system.

- Governance problems for the digital world involve multiple stakeholders. Governments can establish regulation in local affairs, but these regulations may have no effect, because the digital world is global, yet the problems are local. So the digital world may deal with local problems on a global scale.
- Some stakeholders are not territorial. The digital world is global and virtual, and decisions about the digital world that are taken by a government or even by a firm may affect a multitude of people all over the world. Social media platforms, for instance, can have features that affect elections in a country, although decisions regarding the platform's technical architecture are made by headquarters located in another country. For example, the Right To Be Forgotten (RTBF) ruled that search engines are required to remove from the list of search results, when requested by an individual, links to web pages that contain "inadequate, irrelevant or no longer relevant, or excessive" information about that individual. The current RTBF ruling only applies to searches within European versions of the search engines (such as Google and Bing); it does not apply to the US and many other countries (Xue et al. 2016).
- The governance of the digital world is not restricted to governments or firms. Users, represented by civil society organizations—such as NGOs—must be involved in the digital governance process, which should involve collective actions by governments, firms, and users of various digital commons.
- Digital world governance faces an ownership problem. Platforms that store large data are private, with a market for oligopoly construction. The effects of the use of social media and platforms are public, but they are a result of private companies' technological investments and efforts. The digital world relies on multiple transnational communication networks (e.g., global submarine cable net-

works), where multinational companies own and operate the global infrastructure, which ensures connectivity that connects everything: people, devices and data on the global Internet. The private sector has roles and responsibilities in the digital governance process.
- Accountability and compliance mechanisms extend to governments and firms operating in the digital world.
- Digital governance requires a set of comprehensive principles that improve behavior and promote solidarity and inclusion in the digital world through public values. Digital governance requires mechanisms of cooperation and collaboration, with the aim of promoting the integral development of the digital world. Cooperation and collaboration are needed to promote services anchored in ethical and political principles that promote societal development.
- Digital governance requires institutions with authority to govern and build public policies aimed at society's well-being, and to promote human organization's development.
- The digital world has multiple stakeholders and multiple decision centers, containing public and private characteristics. For example, a large part of the digital world's communication infrastructure is private and global.

Looking at these different problems, the concept of digital governance must be comprehensive and holistic to understand the actors, institutions, and processes involved in governance of the digital world. These approaches yield the following definition: *Digital governance is the capacity of polycentric institutions in the digital world to govern (in a legitimate, inclusive, and secure manner) the use of digital commons to produce sustainable services and public policies implemented by governments and firms in a non-territorial and results-based manner.*

The concept of digital governance starts from a set of characteristics that organize means of governing in the digital world. The first feature is that the decision-making process in digital governance must involve multistakeholders and multiple decision centers. Representative democracies have failed to build long-term solutions to the problems of the digital world (Sunstein 2018). Governments have a territorial mandate to solve problems that are global or local. Because problems in the digital world are non-territorial, solutions will not depend only on individual governments, but on multiple stakeholders who are involved in the process of governing the digital world.

These stakeholders include technology companies that deliver systems and applications for use by society, governments that encourage and organize the network structure and regulate its use, users that are subject to these applications and systems, international organizations such as ICANN (of a private nature), or the UN (such as GGE, United Nations Group of Governmental Experts on Advancing Responsible State Behaviour in Cyberspace in the Context of International Security.) and OECD (which create knowledge communities and international agreements).

Multistakeholder processes aim to bring together all major stakeholders in a new form of communication, decision finding (and possibly decision making) on a particular issue. They are based on recognizing the importance of achieving equity and accountability in communication between stakeholders. They involve equitable representation of three or more stakeholder groups and their views and are based on democratic principles of transparency and participation. They aim to develop partnerships and strengthened networks between and among stakeholders (Almeida et al. 2015; Hemmati 2002). There is no single model for the implementation of a multistakeholder approach, but there are several ways in which public problems can be addressed through mechanisms involving the various actors in the decision process (Budish et al. 2015). The multistakeholder is an approach for solving collective action problems and promoting advisory to governments implementing policies (Sahel 2016). It develops through the formation of forums that bring together governments, the private sector, and civil society organizations. The construction of authority stems from the ability to produce consensus among stakeholders, which determine the agenda, workflow, and advisory role of changes and regulations. More specifically, the digital world has multiple decision centers that define standards, norms, and regulations, and that includes international organizations such as the United Nations' International Telecommunication Union (UN-ITU), and private organizations that set standards and recommendations on the use of technology, such as ICANN. The NETmundial conference (The Global Multistakeholder Meeting on the Future of Internet Governance held in São Paulo in 2014) made clear the importance of multistakeholder systems in the context of the global governance of the Internet (Kleinwächter and Almeida 2015).

The discussion on capacity is another central point for digital governance. Capacity discussion is usually associated with the state's performance in society, without a consensus on how these capacities should be allocated and motivated (Cingolani 2013; Jessop 2001). The concept of state capacity is used to understand how the state apparatus produces results for society and the economy through public policies (Matthews 2012). In this perspective, capacity is a stock of skills and abilities that affect policy implementation (Centeno et al. 2017).

The concept of capacity as organizational and institutional stock needs to consider the role of individuals in the context of organizations. In this way, organizations are collective actors whose behavior is reflected in capacity vis-à-vis the organizations of which they are part (Cyert and March 1963). Thus, capacity depends on individuals' actions in the context of complex organizational structures. It requires contemplation on how the institutional framework brings together individuals in the context of organizations (March and Olsen 1984).

The central feature of digital governance arrangements is that it is global, involving a multistakeholder approach. Problems such as the spread of fake news require global solutions. For example, how can a country like Brazil deal with fake news in its elections disseminated by sites and services hosted in other countries? The Brazilian government, alone, has no control over the content generated in another country or efficient ways to contain its dissemination in a network. What is the role of social media platforms such as Twitter, Facebook, WhatsApp, and Instagram to contain disinformation? Note that the problems are often local, but their scale is global. These problems require governance structures with involvement of all actors on a global scale. The governance structure applied to governing the digital world requires capacities in the complex and polycentric power of digital technologies.

Building capacities of governance structures to deal with the tragedy of the digital world means enhancing multistakeholder organizations that address specific organizational issues. The first problem surrounds what information and incentives are provided for individuals to collaborate with the organizations of which they are a part. The second problem concerns the best way to allocate individuals according to their competences within the organization. The third problem hinges on how organizations—comprised of multiple individual and management mechanisms according to contract terms and organizational culture—affect performance (March and Olsen 1985; Simon 1951).

The capacities of digital governance must be activated and motivated to build a human organization ready to solve collective action problems. This requires an ability to organize and build a legitimate polycentric framework, capable of governing the digital world legitimately, inclusively, and securely. These capacities must be activated to implement the elements of governance highlighted in the previous section. Multistakeholders' capacity to govern the digital world in a legitimate, inclusive, and secure manner requires that elements of governance be applied to constitute principles—in the form of ethical and technical standards—along with leadership from the perspective of states and markets, and agency capacity.

Digital governance demands the decision-making process be participatory and deliberative, to generate various agents' collaboration and cooperation. Governments, firms, nonprofit organizations, and users of various services available in the digital world must participate in the decision-making process to promote the legitimacy of the decisions taken. Likewise, digital governance should encourage governments and firms to be accountable and transparent in promoting compliance. Algorithms, processes, and data collection (regardless of whether they are confidential)—as well as the data's treatment—must be transparent, accessible, and direct so that different stakeholders can assess governments' and firms' actions in the digital world. Finally, digital governance structures should encourage coordinated work among different actors. Governments and firms must act in a coordinated way for the production of services and policies guided by the public values of the digital world.

Digital governance must be guided by principles centered on the values of a democratic society (Donahoe and Metzger 2019). The possibility of reducing the respect to privacy, losing accountability, and promoting injustice might destroy trust in the use of digital technologies in governments and firms. Digital governance should aim to promote human benefit by avoiding the tragedy of the digital world. The risks associated with not governing the digital world—because of misappropriation of data and information produced by society—require a global governance framework that is shared in a way that promotes human dignity, democratic accountability, and principles of a free society.

Big Questions

The overarching question regarding the digital world is how can we design a framework of governance that enhances global systemic trust in digital technologies, while simultaneously creating opportunities for all countries to advance their strategic and economic interests in the digital world? To search for answers to this question, we need to examine challenges grouped into two categories: (1) how to use digital technologies to govern better, and (2) how to govern digital technologies—based on pillars such as respect for human rights, international law, and availability of meaningful opportunity—for all people and nations.

How Can We Use Digital Technologies to Govern?

Digital technologies can be valuable tools to create better services and policies that benefit society as a whole, in particular the most vulnerable groups. Mechanisms and procedures are needed to assure that the digitization of government services meets society's needs. Digital government services are usually developed and operated in a decentralized way and distributed over different parts of the public administration. The governance ecosystem of government services has many players with quite different legal statuses, which depend on the government's formal organization as well as the informal power structure, based on political interests (Almeida et al. 2019). As the use of digital technology expands and accelerates in public and private organizations, new questions are posed to society.

- How can society assure the inclusive nature of digital public services, so that no one is left behind?
- How can governments and companies maximize the potential benefits of digital technologies to society?
- Implementation of technology in smart cities must respond directly to citizens' needs. How can governance take care of citizens' interests in digital technology-driven city projects?
- How can governments and corporations minimize new risks, such as the proliferation of disinformation, discrimination of vulnerable groups of society, electoral manipulation, and surveillance?
- How can society assure that governments and companies protect citizens' personal data and privacy?
- How can society ensure the accountability and transparency of digital government actions and policies?

The use of digital technologies can benefit society in many ways. We can have more security, better services, better policies, greater integrity from automation, and ways to share knowledge and information that empower citizens. The relation between human organization and digital technologies has multiple faces. On one hand, governments and private companies use digital technologies for economic, social, and political gains. Digital technologies can improve governance and its elements, promoting greater participation, transparency, and accountability and ways of coordinating the various organizations aiming at the quality of services and public policies. On the other hand, there is a dark side where digital technologies can be appropriated by authoritarian agents, and promote different forms of political and social injustice (Howard 2011). We are at a critical juncture where the relationship between human organizations and the digital world demands the use of digital technologies to improve governance and simultaneously requires new mechanisms to govern the digital world and promote human development.

How Can We Govern Digital Technologies?

Digital technology has been incorporated into every facet of everyday life. While algorithms, IoT, AI, and face recognition technologies provide many benefits and new efficiencies, they also may have negative effects on the protection of human rights. The trend toward algorithmic governance raises many challenges regarding both the private sector and public decision making. Algorithm governance is a component of the digital process; it can vary from the strict legal and regulatory viewpoints to a purely technical standpoint. The inherent opacity and lack of transparency in algorithmic governance may have significant consequences for human agency and accountability.

Artificial intelligence also increasingly dictates public sector governance decisions. For example, predictive algorithms are used by states to calculate future risks posed by inmates, and have been used in sentencing decisions in court trials. Many governments now use facial recognition to exert greater control on society. Algorithms are used in decisions about who has access to public services and who undergoes extra scrutiny by law enforcement. For example, algorithm-based facial recognition allows border control agents to automate passport checks (Carlos-Roca et al. 2018). Risk-assessment algorithms have been used to identify "vulnerable" children and potential victims of child sexual exploitation, so that the government can act to protect them. But the algorithms' opacity makes oversight of these processes difficult. As human agency for decisions is being displaced by AI and other digital technologies, new questions must be asked:

- How can society assure that protection of human rights continues to be a central point of good governance if invisible algorithms dictate governance outcomes?
- How can society assure that algorithmic and AI governance can be reconciled with a human rights governance framework?

- What are the consequences for the protection of human rights when human beings are displaced as agents of governance decision making?
- How can society assure that algorithmic decisions do not create discriminatory or unjust impacts to vulnerable segments of the society?

The difference between a good and bad IoT depends on society's ability to construct effective IoT governance models (Berman and Cerf 2017). By the same token, the process of society's and the economy's digital transformation will be strongly influenced by the digital world's governance framework. Policymakers will need models, structures, and capacity of governance to help their societies and economies prosper in the digital world, so it is necessary to think about the perspective of *governance for digital*. Free societies govern the use of their technologies.

The challenge is to think about a framework that establishes compliance with human rights and institutional norms in a global way, considering that cyberspace is non-territorial. Governments, businesses, civil society organizations, and individuals must adhere to promote further dialogue, greater transparency, and derive a framework of principles and rules that can establish cooperation and solve problems surrounding collective action in the digital world. The next two chapters deal with this perspective. Chapter 3 examines the digital for governance, while Chap. 4 focuses on the governance for digital.

References

Almeida, V., Doneda, D., & Monteiro, M. (2015). Governance Challenges for Internet of Things. *IEEE Internet Computing, 19*(4), 56–59. https://doi.org/10.1109/MIC.2015.86.

Almeida, V., Getschko, D., & Afonso, C. (2015). The Origin and Evolution of Multi-stakeholder Models. *IEEE Internet Computing, 19*(1), 65–69. https://doi.org/10.1109/MIC.2015.15.

Almeida, V., Filgueiras, F., & Gaetani, F. (2019). Principles and Elements of Governance of Digital Public Services. *IEEE Internet Computing, 23*(6), 48–53.

Bannister, F., & Connolly, R. (2018). The Fourth Power: ICT and the Role of the Administrative State in Protecting Democracy. *Information Polity, 23*(3), 307–323. https://doi.org/10.3233/IP-180072.

Barlow, J. P. (1996). *A Declaration of Independence of Cyberspace*. Davos: World Economic Forum. Retrieved from https://www.eff.org/cyberspace-independence.

Bell, S., & Park, A. (2006). The Problematic Metagovernance of Networks: Water Reform in New South Wales. *Journal of Public Policy, 26*, 63–83. https://doi.org/10.1017/S0143814X06000432.

Benkler, Y. (2000). From Consumers to Users: Shifting Deeper Structures of Regulation Toward Sustainable Commons and User Access. *Federal Communications Law Journal, 52*(3), 561–579.

Benkler, Y. (2017). Law, Innovation, and Collaboration in Networked Economy and Society. *Annual Review of Law and Social Science, 13*, 231–250. https://doi.org/10.1146/annurev-lawsocsci-110316-113340.

Berman, F., & Cerf, V. G. (2017). Social and Ethical Behavior of Internet of Things. *Communications of the ACM, 60*(2), 6–7. https://doi.org/10.1145/3036698.

Bindu, N., Sankar, C. P., & Kumar, K. S. (2019). From Conventional Governance to E-Democracy: Tracing the Evolution of E-Governance Research Trends Using Network Analysis Tools. *Government Information Quarterly, 36*(3), 1–15. https://doi.org/10.1016/j.giq.2019.02.005.

Bouckaert, G., Peters, B. G., & Verhoest, K. (2010). *The Coordination of Public Sector Organizations.* London: Palgrave Macmillan.

Bovaird, T., Stoker, G., Jones, T., Loeffler, E., & Pinilla-Rocancio, M. (2015). Activating Collective Co-Production of Public Services: Influencing Citizens to Participate in Complex Governance Mechanisms in the UK. *International Review of Administrative Sciences, 82*(1), 47–68. https://doi.org/10.1177/0020852314566009.

Bowers, J., & Zittrain, J. (2020). Answering Impossible Questions: Content Governance in an Age of Disinformation. *The Harvard Kennedy School (HKS) Misinformation Review, 1*(1), 1–8. https://doi.org/10.37016/mr-2020-005.

Budish, R., Gasser, U., & Myers West, S. (2015). *Multi-stakeholder as Governance Groups: Observation from Case Studies. Berkman Klein Center for Internet and Society at Harvard University.* Berkman Center Research Publication, 2015-1. Retrieved from https://cyber.harvard.edu/publications/2014/internet_governance.

Caplan, R., & Boyd, D. (2016). Who Controls the Public Sphere in an Era of Algorithms? Mediation, Automation, Power. *Data & Society.* Retrieved from https://datasociety.net/pubs/ap/MediationAutomationPower_2016.pdf.

Carlos III University of Madrid. (2018). Study Analyzes the Impact of Targeted Facebook Advertising on U.S. Elections. *Phys.org.* Retrieved from https://phys.org/news/2018-11-impact-facebook-advertising-elections.html.

Carlos-Roca, L. R., Torres, I. H., & Tena, C. F. (2018). *Facial Recognition Application for Border Control.* In: Proceedings of the 2018 International Joint Conference on Neural Networks (IJCNN), Rio de Janeiro, 2018, 1–7.

Centeno, M., Kohli, A., & Yashar, D. (2017). Unpacking States in the Developing World. Capacity, Performance, and Politics. In D. Mistree, M. Centeno,

A. Kohli, & D. Yashar (Eds.), *States in the Developing World* (pp. 1–32). Cambridge: Cambridge University Press. https://doi.org/10.1017/CBO9781316665657.002.

Chen, Y. F. (2013). See You on Facebook: Exploring Influences on Facebook Continuous Usage. *Behaviour & Information Technology, 33*(11), 1208–1218. https://doi.org/10.1080/0144929X.2013.826737.

Chen, Y., & Hsieh, J. (2009). Advancing E-Governance. Comparing Taiwan and United States. *Public Administration Review, 69*(s1), 151–158. https://doi.org/10.1111/j.1540-6210.2009.02103.x.

Cingolani, L. (2013). *The State of State Capacity: A Review of Concepts, Evidence and Measures.* UNU-MERIT Working Paper Series on Institutions and Economic Growth, IPD WP13, #2013-053.

Clark, D. (2010, March 12). *Characterizing Cyberspace: Past, Present and Future.* MIT CSAIL, Version 1.2. Retrieved from https://projects.csail.mit.edu/ecir/wiki/images/7/77/Clark_Characterizing_cyberspace_1-2r.pdf.

Crawford, S., & Ostrom, E. (1995). A Grammar of Institutions. *American Political Science Review, 89*(3), 582–600. https://doi.org/10.2307/2082975.

Cyert, R., & March, J. (1963). *Behavioral Theory of the Firm.* Englewood Cliffs, NJ: Prentice-Hall.

DeNardis, L. (2014). *The Global War for Internet Governance.* New Haven: Yale University Press.

DeNardis, L. (2020). *The Internet in Everything: Freedom and Security in a World with No Off Switch.* New Haven: Yale University Press.

Diakopoulos, N. (2016). Accountability in Algorithmic Decision Making. *Communications of the ACM, 59*(2), 56–62. https://doi.org/10.1145/2844110.

Donahoe, E., & Metzger, M. M. (2019). Artificial Intelligence and Human Rights. *Journal of Democracy, 30*(2), 115–126. https://doi.org/10.1353/jod.2019.0029.

Duff, A. S. (2016). Rating the Revolution: Silicon Valley in Normative Perspective. *Information, Communication & Society, 19*(11), 1605–1621. https://doi.org/10.1080/1369118X.2016.1142594.

Filgueiras, F. (2016). Transparency and Accountability: Principles and Rules for Construction of Publicity. *Journal of Public Affairs, 16*(2), 192–202. https://doi.org/10.1002/pa.1575.

Floridi, L. (2018). Soft Ethics, the Governance of the Digital and the General Data Protection Regulation. *A Philosophical Transactions of the Royal Society, 376*(2133). https://doi.org/10.1098/rsta.2018.0081.

Frenken, K., & Schor, J. (2017). Putting the Sharing Economy into Perspective. *Environmental Innovation and Societal Transitions, 23*, 3–10. https://doi.org/10.1016/j.eist.2017.01.003.

Fung, A. (2015). Putting the Public Back into Governance: the Challenges of Citizen Participation and Its Future. *Public Administration Review, 75*(4), 513–522. https://doi.org/10.1111/puar.12361.

Gasser, U., & Almeida, V. (2017). A Layered Model for AI Governance. *IEEE Internet Computing, 21*(6), 58–62. https://doi.org/10.1109/mic.2017.4180835.

Gillespie, T. (2018). *Custodians of the Internet: Platforms, Content Moderation, and the Hidden Decisions that Shape Social Media*. New Haven: Yale University Press.

Hadfield, G. (2016). *Rules for a Flat World: Why Humans Invented Law and How to Reinvent It for a Complex Global Economy*. Oxford: Oxford University Press.

Hardin, G. (1968). The Tragedy of Commons. *Science, 162*, 1243–1248. https://doi.org/10.1126/science.162.3859.1243.

Heald, D. (2006). Transparency as an Instrumental Value. In C. Hood & D. Heald (Eds.), *Transparency: The Key of Better Governance*. Oxford: Oxford University Press.

Hemmati, M. (2002). *Multi-Stakeholder Processes for Governance and Sustainability. Beyond Deadlock and Conflict*. London: Earthscan Publications.

Hess, C., & Ostrom, E. (2007). *Understanding Knowledge as a Commons. From Theory to Practice*. Cambridge: The MIT Press.

Howard, P. N. (2011). *The Digital Origins of Dictatorship and Democracy. Information Technology and Political Islam*. Oxford: Oxford University Press.

Isaak, J., & Hanna, M. (2018). User Data Privacy: Facebook, Cambridge Analytica, and Privacy Protection. *Computer, 51*(8), 56–59. https://doi.org/10.1109/MC.2018.3191268.

Jessop, B. (2001). Bringing the State Back In (Yet Again): Reviews, Revisions, Rejections, and Redirections. *International Review of Sociology, 11*(2), 149–173. https://doi.org/10.1080/713674035.

Jessop, B. (2002). Governance and Meta-governance: On Reflexivity, Requisite Variety and Requisite Irony. In H. Bang (Ed.), *Governance, Governmentality and Democracy*. Manchester: Manchester University Press.

Kaplan, A., & Haenlein, M. (2010). Users of the World, Unite! The Challenges and Opportunities of Social Media. *Business Horizons, 53*(1), 59–68. https://doi.org/10.1016/j.bushor.2009.09.003.

Kennett, P. (2010). Global Perspectives on Governance. In S. P. Osborne (Ed.), *The New Public Governance? Emerging Perspectives on the Theory and Practice of Public Governance*. Abingdon: Routledge.

Kleinwächter, W., & Almeida, V. (2015). The Internet Governance Ecosystem and the Rainforest. *IEEE Internet Computing, 19*(2), 64–67. https://doi.org/10.1109/MIC.2015.49.

Koppenjan, J., & Klijn, E. H. (2004). *Managing Uncertainty in Networks: A Network Approach to Problem Solving and Decision Making*. London: Routledge.

Lazer, D., et al. (2018). The Science of Fake News. *Science, 359*(6380), 1094–1096. https://doi.org/10.1126/science.aao2998.

Mansell, R. (2013). Employing Digital Crowdsourced Information Resources. Managing the Emerging Information Commons. *International Journal of the Commons, 7*(2), 255–277. https://doi.org/10.18352/ijc.395.

March, J., & Olsen, J. P. (1984). The New Institutionalism: Organizational Factors in Political Life. *American Political Science Review, 78*(3), 734–749. https://doi.org/10.2307/1961840.

March, J., & Olsen, J. P. (1985). *Ambiguity and Choice in Organizations.* Oxford: Oxford University Press.

March, J., & Olsen, J. P. (1989). *Rediscovering Institutions. The Organizational Basis of Politics.* New York: Free Press.

Matthews, F. (2012). Governance and State Capacity. In D. Levi-Faur (Ed.), *The Oxford Handbook of Governance.* Oxford: Oxford University Press.

Mayer-Schönberger, V., & Cukier, K. (2013). *Big Data: A Revolution that Will Transform How We Live, Work and Think.* London: John Murray.

Meijer, A. J. (2015). E-Governance Innovation: Barriers and Strategies. *Government Information Quarterly, 32*, 198–206. https://doi.org/10.1016/j.giq.2015.01.001.

Mills, S., & Waite, C. (2017). From Big Society to Shared Society? Geographies of Social Cohesion and Encounter in the UK's National Citizen Service. *Geografiska Annaler: Series B, Human Geography, 100*(2), 131–148. https://doi.org/10.1080/04353684.2017.1392229.

Mondal, M., Silva, L. A., & Benevenuto, F. (2017). *A Measurement Study of Hate Speech in Social Media.* Proceedings of the 28th ACM Conference on Hypertext and Social Media (pp. 85–94). https://doi.org/10.1145/3078714.3078723.

Mueleman, L. (2008). *Public Management and the Metagovernance of Hierarchies, Networks, and Markets.* Heidelberg: Physica.

Nagle, F. (2018). *The Digital Commons: Tragedy or Opportunity? A Reflection on the 50th Anniversary of Hardin's Tragedy of the Commons.* Harvard Business School, Working Paper, 19-060. https://doi.org/10.2139/ssrn.3301005.

Nye, J. S., & Donahue, J. D. (2000). *Governance in a Globalizing World.* Washington, DC: Brookings Institution Press.

O'Neil, C. (2016). *Weapons of Math Destruction. How Big Data Increases Inequality and Threatens Democracy.* New York: Crown.

Oakley, K. (2010, June 10–11). What Is e-Governance? In *Integrated Project 1: e-Governance Workshop.* Strasbourg. Retrieved from https://www.coe.int/t/dgap/democracy/Activities/GGIS/E-governance/Work_of_egovernance_Committee/Kate_Oakley_eGovernance_en.asp.

OECD. (2001). *Governance in the 21st Century.* Paris: OECD Publishing. https://doi.org/10.1787/9789264189362-en.

OECD. (2014). *Corporate Governance*. Paris: OECD Publishing. https://doi.org/10.1787/20776535.

Olsen, J. P. (2017). *Democratic Accountability, Political Order, and Change. Exploring Accountability Processes in an Era of European Transformation*. Oxford: Oxford University Press.

Osborne, S. (2010). Introduction. In S. P. Osborne (Ed.), *The New Public Governance? Emerging Perspectives on the Theory and Practice of Public Governance*. Abingdon: Routledge.

Osborne, S., Strokosch, K., & Radnor, Z. (2018). Co-Production and Co-Creation of Value in Public Services. A Perspective From Service Management. In T. Brandsen, B. Verschuere, & T. Steen (Eds.), *Co-Production and Co-Creation. Engaging Citizens in Public Service* (pp. 18–26). New York: Routledge.

Ostrom, E. (1990). *Governing the Commons: The Evolution of Institutions for Collective Action*. Cambridge: Cambridge University Press.

Ostrom, E. (2010). Beyond Markets and States: Polycentric Governance of Complex Economic Systems. *American Economic Review, 100*, 1–33.

Pagano, M., & Volpin, P. F. (2005). The Political Economy of Corporate Governance. *American Economic Review, 95*(4), 1005–1030.

Palvia, S., & S. Sharma (2007). E-Government and E-Governance: Definitions/Domain Framework and Status Around the World. *Computer Society of India*. Retrieved from https://www.researchgate.net/publication/268411808_E-Government_and_E-Governance_DefinitionsDomain_Framework_and_Status_around_the_World.

Peters, B. G. (1998). Managing Horizontal Government: The Politics of Coordination. *Public Administration, 76*(2), 295–311. https://doi.org/10.1111/1467-9299.00102.

Peters, B. G. (2004). Governance and Public Bureaucracy: New Forms of Democracy or New Forms of Control? *Asia Pacific Journal of Public Administration, 26*(1), 3–15. https://doi.org/10.1080/23276665.2004.10779282.

Peters, B. G. (2010). Meta-Governance and Public Management. In S. P. Osborne (Ed.), *The New Public Governance? Emerging Perspectives on the Theory and Practice of Public Governance*. Abingdon: Routledge.

Peters, B. G., & Pierre, J. (2000). Citizens Versus the New Public Manager: The Problem of Mutual Empowerment. *Administration & Society, 32*(1), 9–28. https://doi.org/10.1177/00953990022019335.

Peters, B. G., & Pierre, J. (2016). *Comparative Governance: Rediscovering the Functional Dimension of Governing*. Cambridge: Cambridge University Press.

Powell, W. (1990). Neither Market nor Hierarchy: Network Forms of Organization. *Research in Organizational Behaviour, 12*, 295–336.

Powell, W., & DiMaggio, P. (1991). *The New Institutionalism in Organizational Analysis*. Chicago: The University of Chicago Press.

Qiang, X. (2011). Liberation Technology: The Battle for Chinese Internet. *Journal of Democracy,* 22(2), 47–61. https://doi.org/10.1353/jod.2011.0020.

Rahman, S. (2017, October 11). Monopoly Men. *Boston Review: A Political and Literary Forum.* Retrieved from http://bostonreview.net/class-inequality/k-sabeel-rahman-monopoly-men.

Rhodes, R. A. W. (1997). *Understanding Governance. Policy Networks, Governance, Reflexivity, and Accountability.* Buckingham: Open University Press.

Rhodes, R. A. W. (2007). Understanding Governance: Ten Years On. *Organization Studies,* 28(8), 1243–1264. https://doi.org/10.1177/0170840607076586.

Rosenau, J. (1995). Governance in the 21st Century. *Global Governance,* 1(11), 13–43.

Rosenau, J. (1999). Toward an Ontology for Global Governance. In M. Hewson & T. J. Sinclair (Eds.), *Approaches to Global Governance Theory.* State University of New York Press: Albany.

Sahel, J. J. (2016). Multi-Stakeholder Governance: A Necessity and Challenge for Global Governance in the Twenty-First Century. *Journal of Cyber Policy,* 1(2), 157–175. https://doi.org/10.1080/23738871.2016.1241812.

Salamon, L. (2002). *The Tools of Government: A Guide to the New Governance.* New York: Oxford University Press.

Shackelford, S., & Myers, S. (2018). Block-by-Block: Leveraging the Power of Blockchain Technology to Build Trust and Promote Cyber Peace, 19 Yale J.L. & Tech (2018). Retrieved from https://digitalcommons.law.yale.edu/yjolt/vol19/iss1/7.

Simon, H. (1951). A Formal Model of the Employment Relationship. *Econometrica,* 19, 293–305.

Sörensen, E., & Torfing, J. (2005). Democratic Anchorage of Governance Networks. *Scandinavian Political Studies,* 28(3), 195–218. https://doi.org/10.1111/j.1467-9477.2005.00129.x.

Stoker, G. (1997). Governance as Theory: Five Propositions. *International Social Science Journal,* 50(155), 17–27. https://doi.org/10.1111/1468-2451.00106.

Stoker, G. (2018). Can the Governance Paradigm Survive the Rise of Populism? *Policy & Politics,* 47(1), 3–18. https://doi.org/10.1332/030557318X15333033030897.

Stoker, G. (2019). Embracing Complexity: A Framework for Exploring Governance Resources. *Journal of Chinese Governance.,* 4(2), 91–107. https://doi.org/10.1080/23812346.2019.1587859.

Stone, D. (2008). Global Public Policy, Transnational Policy Communities, and Their Networks. *Policy Studies Journal,* 36(1), 19–38. https://doi.org/10.1111/j.1541-0072.2007.00251.x.

Sunstein, C. (2018). *#Republic: Divided Democracy in the Age of Social Media.* Princeton: Princeton University Press.

Sutherland, W., & Jarrahi, M. H. (2018). The Sharing Economy and Digital Platforms: A Review and Research Agenda. *International Journal of Information Management, 43*, 328–341. https://doi.org/10.1016/j.ijinfomgt.2018.07.004.

Tang, G., & Lee, L. F. (2013). Facebook Use and Political Participation: The Impact of Exposure to Shared Political Information, Connections with Public Political Actors, and Network Structural Heterogeneity. *Social Science Computer Review, 31*(6), 763–773. https://doi.org/10.1177/0894439313490625.

United Nations. (2019). *The Age of Digital Interdependency*. Report of the Secretary-General's High-level Panel on Digital Cooperation. New York: United Nations. Retrieved from https://www.un.org/en/pdfs/DigitalCooperation-report-for%20web.pdf.

UNPAN. (2011). *On-Line Glossary on Governance and Public Administration*. United Nations Public Administration Network (UNPAN). Retrieved from http://www.unpan.org/Directories/UNPAGlossary/tabid/928/Default.aspx.

Van Dijk, J. A. G. M. (2012). Digital Democracy: Vision and Reality. In I. Snellen, M. Thaens, & W. Van der Donk (Eds.), *Public Administration in the Information Age: Revisited* (pp. 49–62). Amsterdam: IOS Press.

Vardi, M. (2018). Move Fast and Break Things. *Communications of the ACM, 61*(9), 7–7. https://doi.org/10.1145/3244026.

Warren, M. (2009). Governance-Drive Democratization. *Critical Policy Studies, 3*(1), 3–13. https://doi.org/10.1080/19460170903158040.

Welchman, L. (2015). *Managing Chaos: Digital Governance by Design*. New York: Rosenfeld Media.

Werbach, K. (2002). A Layered Model for Internet Policy. *Journal on Telecommunications and High-Tech Law, 1*, 37.

Williamson, O. E. (2002). The Theory of the Firm as Governance Structure: From Choice to Contract. *Journal of Economic Perspectives, 16*(3), 171–195.

World Bank. (1997). *World Development Report 1997: The State in a Changing World*. Washington, DC: World Bank. Retrieved from http://documents.worldbank.org/curated/en/518341468315316376/World-development-report-1997-the-state-in-a-changing-world.

Xue, M., Magno, G., Cunha, E., Almeida, V., & Ross, K.W. (2016). *The Right to Be Forgotten in the Media: A Data-Driven Study*. In: Proceedings on Privacy Enhancing Technologies, 2299.

Zuboff, S. (2019). *The Age of Surveillance Capitalism: The Fight for a Human Future at the New Frontier of Power*. New York: Public Affairs.

CHAPTER 3

Digital Technologies for Governance

Abstract The use of digital technologies in governance processes has been adopted by both public and private sectors. Digital tools can be applied to promote governance and expand public service delivery and public policy capacities. Artificial intelligence and algorithmic governance increasingly dictate public sector governance decisions. This chapter explores how digital technologies have been used by public administration to implement public policies and services. It also discusses how the use of digital tools for governance entails a new set of risks for governments and firms.

Keywords Tools • Algorithms • Artificial intelligence • Public services • Organizations • Risks • Public policies • IoT • Government • Platform

The focus of this chapter is to look at the use of digital technologies in governance processes in both public and private sectors. Here we examine the virtues and problems of using digital technologies for governing the design, implementation, and operation of services and policies. The use of digital technologies can be applied in governing to construct policy tools. As a policy tool, the application of technologies is conditional on managers' choices and how they catalyze changes in public and private organizations. Technology is not the only decisive factor for the construction of

© The Author(s) 2021
F. Filgueiras, V. Almeida, *Governance for the Digital World*,
https://doi.org/10.1007/978-3-030-55248-0_3

digital governance. Digital governance depends on the choices made and how they are absorbed within organizations; and it could be viewed as a new way to think about how collective action is organized.

In addition to addressing digital technologies as tools for the implementation of policies and services, we discuss how technologies transform organizations, modifying basic governance structures, creating new modes of interaction between the state, business, and society. Finally, we address the new risks associated with this process of transforming the human organization and why governing digital technologies is fundamental for contemporary societies.

The Centrality of Algorithms

Algorithms are sequences of instructions or steps for solving a particular problem or task. They have been put to work by human beings in all kinds of decision-making processes for centuries. Using algorithms in governance is not new (e.g., predetermined sets of rules for problem solving and pre-established routines for government decision making). Since the advent of "big data analytics," a wide spectrum of different types of algorithms have been deployed to guide data analysis and governance outcomes in society, ranging from routines that are fully programmed by human beings to AI systems that learn from big data and autonomously make governance decisions.

A class of algorithms (called *machine learning algorithms*) have been developed to tackle complex problems, through fast sorting, analysis, and pattern recognition applied to massive quantities of data that are beyond the scope of human capacity to analyze (Domingos 2015). Some examples include image identification, speech recognition, spam identification, driving, and human behavior prediction.

In 1959, Arthur Samuel defined machine learning as a "field of study that gives computers the ability to learn without being explicitly programmed." Machine learning algorithms are programs that "learn from experience" and exposure to data (Samuel 1959). Machine learning algorithms achieve good results when "supervised" or "trained" on a predefined dataset that incorporates training examples. In addition, "unsupervised" machine learning algorithms can find patterns and relationships in datasets related to a specific problem without being trained. With "bigger" and even more "generalized" datasets, machine learning becomes less dependent on human-coded algorithms that direct outcomes, and moves toward the goal of artificial general intelligence. The term artificial intelligence, as defined by Herbert Simon, deals in particular with the phenomena that appear when computers perform tasks that, if performed by people, would be regarded as requiring intelligent thinking (Simon 1969). Machine learning algorithms are a subfield of AI.

For example, the application of mobile phone tracking during the SARS-COV2 pandemic in China sought ways of monitoring and surveillance of social detachment to prevent rapid contagion and flattening the contamination curve. Such measures can do injustice and create exclusion zones. The results of government actions like this depend on the entire chain of commands—based on the input data—to constitute the solution to the proposed problem (Servick 2020). A concern here is how impartial this programmed logic actually is, and whether it respects human rights and the quality of input data, so that it can have effective and real solutions. Suddenly, the algorithms are central to the policy process. Likewise, in all digital transformation, algorithms are central to governance. Algorithms make solutions real, even if partial or with governance failures. For this reason, they represent the important topic of all-digital governance.

Algorithms and intelligent machines already guide a vast array of technology-driven outcomes in both the private and public sectors. For example, private sector digital platforms such as Google, Facebook, WhatsApp, Twitter, YouTube, Uber, and Airbnb effectively control global access to information and services. They play a significant role in setting the parameters of freedom of expression through their proprietary algorithms. The power and effect of these algorithms is not conspicuous or discernible to most users. The use of algorithms in public and private decision-making processes has given rise to the concept of algorithmic governance, where algorithms assist, supplement, or in many cases, replace human decision making.

Artificial intelligence and algorithmic governance also increasingly dictate public sector governance decisions. For example, predictive algorithms are used by states to calculate future risks posed by inmates, and have been used in sentencing decisions in court trials. Algorithms can help make the judicial and legal justice processes more efficient and less expensive, and even reduce prison crowding. But replacing human judgment in governance decisions with algorithmic governance has not been considered or adequately theorized.

Using digital technologies in governance processes is related to the ability of tools to shape and guide the behavior of bureaucrats and citizens, and to delimit collective solutions to public problems. For example, the interconnected network of data collection devices provides a fundamental tool for improving public policies and public services (Greengard 2015; Danaher et al. 2017). Public service administrations aim at automating service provisions and making use of massive distributed data and algorithms to anticipate future demands (Williamson 2014; Chen and Hsieh 2014). The benefits of digital public services are various (Almeida et al. 2019). One benefit is the participation of society in the design and implementation of innovative public services. As an example, smart services require citizens and firms to act as co-producers of public services (Williamson 2014). In addition, algorithms provide the ability to produce predictions about individuals' behavior and use this predictive ability to alter expectations and induce incentives for service users (Dunleavy 2016). In short, public services have the potential to be fast, cut red tape, and be more accountable and effective with the use of digital technologies.

In public policies, algorithmic decision systems can be used by governments to change the social order in different areas, such as education, health, criminal justice systems, and social policies (Gillingham 2015). Digital tools can change the social order by analyzing the incentives given to various policy actors and that can be predicted by algorithms, fulfilling important policy functions (Danaher 2016). Algorithms can filter and direct information to change citizens' perceptions. And changes in citizens' perceptions can create a context of greater or lesser confidence in the functioning of institutions. Algorithms affect culture, modify knowledge, and build realities (Just and Latzer 2017; Napoli 2014).

Mobile phone tracking in China, using the WeChat application during the SARS-COV2 pandemic, may have fulfilled an important social function (Boulos and Geraghty 2020). It made it possible to monitor and maintain citizens' social distance to ease the demand from the health system and cross the peak contagion, as well as making social aid payments during quarantine. This algorithm was based on important prerequisites, such as the extension and use of WeChat by almost all Chinese citizens, its existing payment functions and the quality of the data collected. However, the WeChat algorithm now has a permanent surveillance and control function, expanding governance capacity, but breaking fundamental rights such as privacy. Evidently China's political conditions are totally different from the Western world. But the WeChat algorithm, for example, can be expanded to authoritarian and unlawful uses, changing the political culture and functions of institutions.

Algorithms play an important role in promoting the coordination of public policies and services. That is, algorithms have the power to predict and encourage behavior, solve problems of coordination, and promote collective action. In addition, algorithms—when operating massive data—are able to promote actors' behavior patterns and serve as a tool for accountability and transparency. Finally, they can make decisions that affect society (König 2019). In the digital world, algorithms are essential for defining technology's role, considering the process of selecting, classifying, and defining citizens' rights and benefits. The Social Credit System in China relies on digital technology to construct a set of mechanisms providing rewards or punishments as feedback to actors, based on many signals that represent the morality of their actions, covering economic, social, and political conduct (Creemers 2018). Algorithms are a new kind of ecosystem, subject to pollution, threats, and exploitation (Domingos 2015). If in the traditional human-based public service, street-level bureaucrats have the role of establishing the selection of rights and benefits (Lipsky 2010), then in the digital world, human interactions are being gradually replaced by algorithms. In this way, algorithms can reproduce human procedures and present the same gaps between the objective of public policy and the decision made. With the digital technologies for governance enhancement, we observe a move from street-level bureaucrats to street-level algorithms. Street-level algorithms are layers of systems that interact with citizens. Street-level algorithms are responsible for making decisions that affect the users of public policy and services and their stakeholders (Alkhatib and Bernstein 2019).

For example, the digitalization of the Cadastro Único (Unified Register) for social programs in Brazil and the collection and interoperability of different government databases has enabled greater controls over the accuracy of payments and expanded various uses in other public policies. However, the algorithm that makes the selection of beneficiaries increased the time for granting the benefit, excluded several beneficiaries, and reduced the scope of important cash transfer policies. Likewise, the digitization of public pension benefits created greater difficulties for this public service, making it harder to grant benefits, worsening digital service, and increasing red tape and citizens' dissatisfaction.

The algorithm is central because it contains the design of policies and services. Within the design of the algorithms and their programming is where policy choices are made. The algorithm emulates the policy or service process, encodes that process, and produces—from the input data—the proposed solutions.

By removing human interactions, the use of algorithms for digitizing public services and public policy becomes central to governance processes. Algorithms embody the characteristics of the design of public policies and services (Janssen and Kuk 2016). Algorithms deal with public policy problems. They are able to represent design solutions without human interaction. They can be used for nudging users and influencing the behavior and decision of individuals and groups (Kosters and Van der Heijden 2015). Algorithms should not be viewed only within the governance process. They exercise a technocratic authority that has political consequences. In making a decision, algorithms first decide on an epistemic bias. As a technology of power, algorithms technically decide between governing and governed bodies (Janssen and Kuk 2016; Danaher 2016). Important decisions are left to algorithms' discretion (Andrews 2018). They can change fundamental aspects of bureaucracy and the way the government relates to society. That is why algorithms are central to understanding the application of digital tools to governance.

Digital governance facilitates the implementation processes and contributes to the improvement of government actions (Milakovich 2012). Digital governance tends to reinforce a technological determinism that leads us to easy solutions to problems that are wicked (Kitchin 2014). A technocratic perspective of digital governance assumes that social problems can be resolved simply by using digital technologies. However, it is evident that the use of digital technologies in public policies is not a panacea or a definitive solution to various public problems (Alkhatib and Bernstein 2019).

Digital Tools

The use of digital technologies can improve governance, and their adoption by public administration should focus on tools and instruments for implementing public policies and services. The concept of public service is a modern concept that means the structure of services implemented by governments to benefit the community and ensure rights (Duguit 1923). On the other hand, public policy is what the government chooses to do (or not do) to implement actions in society (Howlett 2014; Dye 1972). Public service governance involves choosing tools to implement rights and benefits for users of this service structure. The governance of public policies involves the choice of tools to implement government actions in society, changing the social structure.

Policy tools are "techniques of governance that, one way or another, involve the use of state authority or its conscious limitation" (Howlett 2005, 31). Policy tools express the exercise of state authority and delimit the behavior of individuals in society. Policy tools are technologies of power, which result in the normative orientation of individual behavior and produce results that are expected by the governing. Traditional examples of policy tools in the literature are subsidies, regulation, legislation, budget, standards, guidelines, grants, programs, taxes, and public ownership.

Tools are central to the governance process. The governance paradigm has given greater centrality to tools. While a traditional public administration is based on the proposition of agency and programs, governance is based on the effective use of tools to improve the policy-implementation process (Salamon 2001). The use of governance tools contributes to solving problems associated with *agencification*, which refers to the creation of semi-autonomous public organizations that are independent from government (Moynihan 2006). Agencification promotes fragmentation of public services and governance failures, while the use of tools enables reintegration of services and greater capabilities to address policy implementation challenges. For example, the digitization of public services prevents citizens from having to go from agency to agency for a benefit or right. The use of digital tools reintegrates services in one place, promoting greater efficiency.

Policy tools are chosen by managers to achieve policy goals. The choice of tools depends on factors such as context and time, and they impact the coherence, consistency, and congruence of policy design. Policies usually are created and implemented through mixes of policy tools (Howlett 2019). Policy design for the digital world must consider policy mixes to produce effective interventions, using digital tools to produce more effective public policies and services. However, choosing the tool is not just a technical issue. It is also political. The use of policy tools delimits the discretion of implementation actors, offers action perspectives, and outlines how policies will be executed. Depending on the choice of tool, typically the delimited actors have more or less discretion and power, as well as the limits of actions of the bureaucracy. Choosing the policy tool creates a battle not only for the most efficient way to solve a problem, but also for the influence that various affected interests will have on policy implementation (Salamon 2001; Peters 2005).

More than a complex set of systems and codes, digital technologies can be understood as tools that mediate state authority to implement public policies and services (Hood and Margetts 2007). In other words, this set of complex systems and algorithms exert state authority to achieve a proposed goal of solving a public problem. Digital tools are policy tools built using digital technologies. They seek to achieve certain goals to change society's behavior, establish forms of regulation, and produce benefits and rights for individuals in society. Digital tools are digital technologies chosen by public managers to implement public policies and services.

Digital tools are subject to technical and political choices, comprising a complex mix that explains a policy implementation's degree of effectiveness (Filgueiras et al. 2019). Digital tools specify technological solutions to be applied to solve public problems. The choice of techniques is based on factors that optimize the production of results for public policies. In this case, digital tools should aim to promote: (1) faster service delivery and policy implementation, as computers are able to analyze millions of individual situations in a short period; (2) lower costs, because systems can perform basic and repetitive tasks with few resources employed; (3) accuracy, as computers can handle complex situations and provide reliable results; and (4) new capabilities, as some applications are impossible to reach without introducing computers.

Currently, technical choices are driven by these four rationale criteria and determined by the available technology. However, even the notion that we should prioritize what technology is available explains why digital tool adoption easily leads to failure, because it disregards institutional factors such as agents' preferences. Political choices surrounding digital tools are made in contexts of uncertainty and bounded rationality, and such choices depend on institutions' capacity to function (Fountain 2004). Even if a digital tool is available, it will not necessarily be chosen or adopted by governments (Balutis 2001; Bannister and Connolly 2011). The preferences are driven by broader institutional factors. These preferences face some bureaucratic barriers such as institutional actors' resistance to change, legislative difficulties, inner characteristics of the services provided, personal preferences, and features of the target audience. Digital tools are embedded within institutional perspectives, for which actors' preferences matter when digitizing and reviewing services and public policies.

The literature on the uses of technologies in government has advanced to a perspective of digital transformation (Bertot et al. 2016). The perspective of addressing the issue of technologies in government has advanced to consider the institutional factors and the role that relevant actors have in the construction of policies. From this perspective, the adoption of technologies in public services and public policies is a more complex process, considering institutional factors (Tassabehji et al. 2016). Technologies' availability interplays with broader institutional factors that explain the success or failure of digital transformation (Weerakkody et al. 2016).

Public Services' Digitization and Platformization
Some countries are adopting platforms that accommodate the digitization process of public services. Digitization means redesigning the public service step by step to transfer these redesigned services to the machine. Digital public services change the interaction between governments and citizens and between governments and firms.

The experience of countries such as the United Kingdom and Brazil with the creation of platforms such as GOV.UK and GOV.BR, respectively, modify these interactions and facilitate the access and use of public services by citizens and companies. These platforms collect user data and apply layers of artificial intelligence to produce service customization and ensure that the user can use the services unassisted.

Public service platformization begins with the choices of managers. They arise as a result of fiscal crises and are embedded in austerity policies that justify the adoption of these digital tools. In the case of the United Kingdom, the process of public service platformization began in the 2008 crisis, creating a large front for the government to digitize public services. Brazil later adopted this process. Only with the deepening of the fiscal crisis in 2016 did initiatives begin to digitize and construct the public services platform.

In both instances, digital public services already existed, but they were outnumbered and scattered. The rationale for public services is to scale the number of digital services, reintegrate them in one place, and redesign them to reduce costs while increasing the services' affordability.

Digital tools for governance seek to build new capacities to deliver digital-based policies and services. The digital governance perspective involves thinking about how digital tools can facilitate and enhance governance processes and strengthen their different elements. The following are key digital tools:

- *Artificial intelligence.* Artificial intelligence is not a simple technology, but rather a set of techniques that simulate human intelligence to support autonomous system decisions. There is no precise definition for AI, and it is not just about the automation of repetitive actions (Gasser and Almeida 2017). Artificial intelligence depends on machine learning ability and uses techniques such as deep learning, rule-based systems, natural language processing, neural networks, and speech recognition to make decisions (Eggers et al. 2017). There are several possible applications for AI in public policy. For example, AI has been applied to health regulation and restaurant inspection (Kang et al. 2013). The numerous possibilities for applying AI in public policies tend toward a growing autonomy of the administrative machine, which decides benefits and rights, as well as punishment and constraints.
- *The Internet of Things (IoT).* The IoT fuses the Internet with the physical world (Weber 2013). It means the use of connected devices that collect and share data and information about various aspects of life for use in public policy and business. The IoT is about a network of networks of uniquely identifiable endpoints or "things" that capture and share data (Nord et al. 2019). The Internet assumes that people create data. The IoT assumes that things create data (Madakam et al. 2015). The IoT applies to industrial facilities, financial services, smart cities, smart energy, connected cars, smart agriculture, public infrastructure (such as sensors in bridges, buildings, and roads), and connected health, among others (Bartje 2016). It also leverages the production and use of data by governments to formulate and implement public policies, such as traffic flow control, and more complex issues such as agricultural production and product tracking.
- *Government as platforms.* Governments have been following a trend toward "platformization" (Helmond 2015). Social media models to promote interaction between two or more distinct parts using platforms. They have a computational aspect, so they must produce

innovative forms of interaction; there is also the political aspect, such as a speech space; the figurative aspect, in which opportunity is an abstract promise as a practical one; and the architectural aspect, in which design matters for construction of interactions (Gillespie 2010). Platforms provide a new mode of communication between governments and society, especially with regard to public services and policies. Platforms represent the natural progress of digital technologies to solve collective problems, where citizens have the skills needed to solve local and national problems, governments provide information and services required by citizens, and citizens are empowered to spark innovation that will improve governance (O'Reilly 2010).

- *Blockchain.* Blockchain is a digital technology based on transaction logging protocols checked in a decentralized manner, dispensing with intermediaries. The blockchain application occurs especially in the financial area. The blocks are stored in a decentralized manner, with well-defined operating protocols (Shackelford and Myers 2018).

These digital tools can be applied to promote governance and expand public service delivery and public policy capacities (Veale and Brass 2019). They enable new capacities and the expansion of possibilities of state action and firms, transforming the behavior of individuals in society. Digital tools transform policies as well as the organizational and political aspects of public policy. It is this disruptive process of organizations that digital governance must address in a manner that can harness the potential of digital technologies in the governance process.

Regarding transparency and accountability, digital tools have brought hope for more open and transparent governments. The assumption is that digital tools can provide citizens with more information and enable greater opportunities for individuals to examine governments and their performance (Margetts 2013). Digital tools can contribute to enhancing political participation in decision making and facilitate transparency and accountability (Ahn and Bretschneider 2011).

Open government initiatives unite transparency and data sharing to achieve social values and participatory governance (Attard et al. 2015). Data sharing increases transparency initiatives, reducing information deficits and providing societal initiatives to control government and create new public policy solutions via greater collaboration between government and society (McDermott 2010). Collaborative solutions are driven by technology creation and new opportunities for society (Chun et al. 2010). The effects of digital tools on public policy is not just to promote automation and greater efficiency, but to create collaborative networks that drive the digital tools and create new capabilities (Mikhaylov et al. 2018).

The possibilities engendered by digital tools to promote transparency are vast. However, the mere notion of transparency lacks a substrate for the problems of information production and the link to accountability. The use of digital tools does not ensure that transparency leads to accountability, because of the political, institutional, and structural variables that organize and produce information. Transparency, enhanced by Internet usage, does not necessarily result in accountable and responsive governments, as it may increase the degree of uncertainty in the scope of governance perspective. This is because the greater availability of data and information on the Internet is a necessary but insufficient condition for society to operate within to learn about the actions of the various agents in the public arena (Lindstedt and Naurin 2010).

Open government initiatives face adoption problems and are surrounded by simplistic views (Janssen et al. 2012). Furthermore, disseminating information to society does not mean that the audience is able to process it, indicating a problem of cognitive mediation of information (Heald 2006). Cognitive mediation for government transparency provides instrumental use of information, providing an opening process buoyed by emphatic discourses. By considering the citizen as a mere consumer of information, transparency through digital tools can fail to produce an understanding of public issues, because it does not allow for an inquiry into the information-construction process (Etzioni 2010). The use of digital tools requires a regulatory apparatus for information production and dissemination, to ensure privacy and information security.

In addition to building open government initiatives to promote transparency and accountability, digital tools facilitate public policy coordination. The use of digital tools makes it possible to establish monitoring and evaluation systems with abundant data on public policies' performance and governance process. Big data and AI, associated with the digitization of public services, make it possible to coordinate the work of different agencies during the policy process. Digitally based monitoring instruments can contribute to horizontal and vertical coordination by allowing the management of large volumes of information by different agents. Digital government platforms promote service reintegration, avoid agencification, and facilitate the delivery of public services to citizens (Dunleavy et al. 2006).

Digital tools do mitigate the effects of noncooperation. Coordination of public policies implies collective action, requiring the engagement of various stakeholders, agents, and cross-sector organizations. Any conflicts caused by differences in data literacy and analytical capabilities of agents may lead to noncooperation. Thus, creating cooperation and collaboration structures in digital governance involves institutional and management arrangements that facilitate interorganizational collaboration. Collaboration, which is essential to coordination processes, requires trust among participants and overcoming problems of the digital divide (Luna-Reyes et al. 2007).

Digital tools can increase citizens' participation in the governance process. Digital tools make other forms of democracy mediated by computers. Digital participation creates opportunities for citizens to participate in the formulation of public policy (Brewer et al. 2006). It can be centered on the governance structure in which connected citizens participate in the policy decision-making process through digital tools such as online forums, virtual public hearings, or the use of wiki technologies (Janssen and Kies 2005). Digital tools for participation are society-centered, promoting activism and the existence of counter-publics (Dahlberg 2007). Digital participation structures can challenge governance structures because they promote activism against state colonization and corporate interests (Jordan 2007).

The use of digital tools is a new form of communication between governments and citizens. However, it presents mixed results. The use of digital tools does not ensure political participation improvement. Communication between state and society has not improved with the advent of social media, because of weak regulatory processes. The spread of fake news and post-truth contexts directly affect electoral processes, worsening the process of public communication (Allcott and Gentzkow 2017). Additionally, it affects the governance of public policies, because of misinformation's influence on political life (Perl et al. 2018). Digital participation structures may affect governance structures without qualifying state action in society and without necessarily influencing the decision-making process, because of the instability and insecurity in the online information ecosystem.

Overall, digital governance means using digital tools to improve governance. Digital tools can create new communication channels between governments, firms, and society. They can promote better policy and service coordination, increase policy efficiency and effectiveness, and decrease spending on repetitive bureaucratic activities. Digital tools foster change, with potential for improving public action while building the state's trust and legitimacy. However, digital governance also has disruptive potential, promoting changes with great possibilities for turbulence in public organizations. Digital tools and their use in governance processes trigger organizational and institutional changes inside governments.

Changing Organizations

The disruptive potential that digital tools provide for organizations is enormous. In recent years, the process of organizational change has emerged from the perspective of digital transformation. Digital transformation is an important phenomenon for research on organizations (Luna-Reyes and Gil-Garcia 2014) as well as practitioners (Westerman 2016). Digital transformation considers that technology is part of a complex puzzle involving organizations, institutions, citizens, and firms to change the organizational value chain.

Governance must consider how its structure's organizational and institutional aspects function in relation to society (Olsen 2010; Stoker 2019). Digital transformation changes governance structures. It transforms organizations by modifying the attributes of the service and management structure as well as their political aspects. With respect to the service structure, digital transformation changes the interaction between citizens and governments and between firms and governments, making this relationship mediated by machines. Regarding the management structure, the digital transformation modifies the procedures and work performed by managers, constituting new capacities. Finally, regarding political aspects, digital transformation changes the interaction between government and society, creating other forms of communication.

Digital technologies play an important role in changing organizations, shaking the governance structure, and making it realized by digital tools. Now the interaction between the organizational aspects of governance as a whole and digital technologies change management practices, the political action of governments, and the relationship between governments and society. Digital governance emerges as a reciprocal relationship between the change of organizations and digital technologies. In other words, technology does not determine organizational change and organizational change does not determine technological change. The relationship between technologies and organizational changes is complex, based on many situations of bounded rationality, ambiguity, and different organizational streams (Cohen et al. 1972).

Conceptually, digital transformation is the process where organizations respond to changes happening in their environment by using digital technologies to alter their value-creation processes (Vial 2019). However, it is a concept still in dispute, and applies to both public and private organizations. Being in dispute, there is still no precision of the phenomenon and its results for society. Digital transformation encompasses both process digitization focused on efficiency, and digital innovation focused on enhancing existing physical products with digital capabilities (Berghaus and Back 2016). There is a prevalence of technological determinism in the definition of digital transformation, which describes changes as imposed by the digital industry to automate various tasks (Legner et al. 2017).

Technology is a necessary but insufficient condition to explain digital transformation. It also depends on how digital tools are absorbed in governance processes, as they change organizations. Understanding this process means designing a theoretical framework that considers the available technology, the choice and use of digital tools, complex organizational structures, and policy work performed by public managers (Sandoval-Almazán et al. 2017). The goal of digital governance is to create public value, rebuild citizens' trust in the functioning of institutions, and provide legitimate forms of management. Digital tools are important, and the digitization of services and public policy is a central topic in digital transformation processes. However, it is necessary to understand the social constructions, behavior, attitudes, and cognitions of individual actors on transformational change (Mergel et al. 2019).

Increased availability of digital tools transforms organizations in their ability to promote public value and effectiveness in service delivery and policy implementation. Depending on the choices that organizations make, digital tools can be used or not and impact public governance in various paths. The use of digital tools provides a means for organizations to tackle disruption, encourage competition, and incentivize issues. Digital tools have led to organizational shifts, stemming from changes in civil society and the political and institutional positioning of public managers. Digital governance can promote services' reintegration into platforms and process reengineering. It can create more agile governments that respond promptly and flexibly to changes in the social environment, because of pressures such as needs-based holism and digitization to drive productivity gains (Dunleavy et al. 2006). Service reintegration, needs-based holism, and digitization are all factors for changing governance (Dunleavy and Margetts 2013). These are the core change inductors for digital transformation:

- *Co-production of services.* The use of digital tools has the potential to promote co production and customization of services. Co-production of services is the involvement of users with the production of services (Linders 2012). This user involvement makes deliveries dependent on process reengineering, so that services become citizen-centric and not just the bureaucracy to the machine. Co-production is necessary so that public services can expand their delivery capacity and anticipate or predict citizens' needs (Meijer 2012). Service digitization strategies should consider co-production as a necessary condition for the increase of efficiency and agility, associated with algorithms that can be used to collect and analyze user data. The objective is to promote increasingly personalized and anticipated public services. Algorithms can collect data and anticipate future individual behaviors (Williamson 2014).
- *Collaboration.* The use of digital tools encourages collaboration between different organizations. The construction of digital tools in the private sector provides cross-functional solutions across business units (Seo 2017; Earley 2014; Maedche 2016). In the public sector, digital tools should encourage collaboration between organizations and prevent the formation of silos (Chen 2017). Collaboration between organizations facilitate policy coordination and new implementation ideas (Agranoff and McGuire 2004).
- *Inclusion.* Public services and public policies produce benefits and rights. Thus, an inclusive perspective must drive them. The use of digital tools by organizations should consider the digital divide as a central element for digital governance. Interaction between individuals and machines requires skills to access economic, political, and social dimensions. The digital divide refers to the growing gap between non-privileged portions of society and increasingly digitized and internet-based organizations. Internet access is a structural condition of the digital divide (Ferro et al. 2011). Another factor of exclusion is digital literacy; this is the individual's skills to use digital technologies to find, evaluate, create, and communicate information

requiring both cognitive and technical skills (Scheerder et al. 2017). Digital governance requires policies to address the digital divide and promote digital literacy, to promote inclusion and greater citizen connectivity. In addition, it should promote user experiences that facilitate interaction between citizens and machines and make services more usable.
- *Data protection and security.* Organizations should promote data protection and security mechanisms that bolster (rather than undermine) citizens' privacy. Data protection depends on regulatory mechanisms and security systems that make it possible to maintain privacy and see which data produces public benefits (Chen and Zhao 2012). Users must be confident in digital tools' goal of building trust, as well as believing in the tools' potential to promote the development of human organization (Wu et al. 2018).

For example, the OECD has encouraged members to promote the digital transformation process of governments and firms.[1] Toolkits were developed to facilitate this process in order to create a knowledge network that underlies the process of change. In the same way, practices and principles were created to define and organize the digital transformation process, to conceive an integrated and collaborative perspective of change. Toolkits aimed at member countries are based on constant service co-production and collaboration processes, to expand the expected impact of digital transformation.

Institutionally, these principles must comply with a set of rules that guide the change of organizations. Digital tools provide new ways for organizations to act and configure new ways of governance. The changes in organizations are neither linear nor are their consequences fully understood. The process of change causes turbulence for organizations. The use of digital tools has disruptive potential for state bureaucracies, altering their relationship with society. The changes depend on factors internal to organizations, as well as the broader political context. Turbulent contexts are more prevalent in the contemporary world. In turbulent times, political regimes seek a new order (Olsen 2007). Turbulence is "a situation where events, demands and support interact and change in highly variable, inconsistent and unexpected and unpredictable ways" (Ansell and Trondal 2018, 44–45).

[1] See https://www.oecd.org/going-digital/.

Digital tools promote turbulence because they change service processes and policies in unexpected ways. Turbulence does not mean to achieve change automatically. The context of turbulence means that the application of digital tools in policies and services favors the disruption of the state's relationship with society. Turbulence can create pressure for organizational solutions and institutional arrangements to be maintained, because governance systems can reinforce practices and formats, reinforcing institutional path dependence (Olsen 2010). Citizens require more than digital tools' availability and use to rebuild confidence in their perception of how well the organization functions (Janssen et al. 2018).

Digital tools transform organizations. They are adopted in institutional contexts where choices are made based on rules, roles, and problems. The logic of appropriateness may be an adequate lens to examine the adoption of technologies by organizations. The logic of appropriateness is a view of action that involves the matching of situations, roles, and rules for decision making (March and Olsen 1989). Appropriateness in bureaucratic organizations defines the basis of decision-making bias according to norms and the context of interpretation performed by the different agents. The logic of appropriateness occurs when a decision is made around social norms that are correct rather than a cost-benefit logic that is considered better. In the logic of appropriateness, choices are not made consequentially, based on the aggregation of preferences of rational actors (March and Olsen 1989).

Initially, the appropriateness of digital technologies underscores a supposed consequentialist logic based on the benefits of technologies. However, the appropriateness of digital technologies occurs in institutional contexts that depend on negotiation and coalitions to promote change based on principles and norms that frame the digital transformation. The appropriateness of digital transformation occurs when there is a perception of low regulation of the digital world by the members of polity (Ciborra 2004). Digital transformation's decision-making process focuses more on consequentialist logic than appropriateness, modifying the general terms of the order and predictability of managers' behavior. Because of that, the digital transformation and adoption of digital tools is disruptive and organizational turbulence ensues.

In many cases of countries that have adopted broad processes of digital transformation, the initial motivations were the fiscal crises. Digital services are less expensive and can bring great benefits. For example, Uruguay facilitated the process of digital transformation with a unique identity for all citizens, after cow fever caused a meat crisis. The United Kingdom responded to the 2008 financial crisis with a deep project to transform and platforming public services, which changed the government's structure (Dunleavy and Margetts 2013). Some countries, such as Estonia, have a longer-lived digital transformation process, because of economic crises and the country's image as a tax haven (Tammpuu and Masso 2018). In these cases, contradictory technologies brought turbulence, which was mitigated by the increasing correction of the use and extension of technologies in public life.

The turbulence caused by adopting digital tools stems from infrastructure issues, with managers and staff learning about new users' relationships, along with the presence of new technologies, cybersecurity issues, legal barriers, ethical issues, and change management. There are two strategies that organizations undertake to deal with turbulence: they stabilize and reproduce patterns of path dependence; or they adapt by producing institutional syncretism, activating experimental fashion management, encouraging flexible and decentralized structures, and developing informal networks and other arrangements to coordinate change (Ansell and Trondal 2018). Digital governance works with a process of digital transformation that requires constant adaptation by organizations.

International experience shows that the preferred strategy is to adapt to the presence of digital tools (Hansen et al. 2011). Digital transformation provides a context of difficulties to plan the future and precludes anticipation. Digital tools' advancement is faster than organizations' capacity to stay apace. The future in the context of digital transformation is difficult to anticipate (Dawes 2009), making adaptation through institutional resilience the strategy adopted by organizations to cope with turbulence (Wildavsky 1988).

The change in organizations due to digital tools is disruptive, because the presence of digital tools is increasing in governance arrangements. Organizations adapt to the process of digital transformation, activating their resilience to maintain their functions and powers with the digital world's growing presence. The digital governance perspective means activating a pattern of organizational change in which actors build organizational flexibility and institutional arrangements to absorb the digital world's complexity and incorporate its principles and tools.

The technocratic temptation of digital governance can widen the scope for risks of inefficiency, ineffectiveness, and exclusion, as well as political problems that contaminate the administration's external environment.

Therefore, it is crucial to understand the nature of the risks and unintended consequences associated with the use of digital tools in governance processes.

Risks

Using digital tools to improve governance entails a new set of risks for governments and firms. Risk management is the capacity with which various actors can anticipate through rational calculation future problems, and control the costs and benefits of action. States and societies employ efforts to calculate and control risks in various aspects of life. They range from individual risks—calculated and monetized by insurance companies—to risks to global societies, such as terrorism or climate change. The fact is that the contemporary world is a risk society constantly trying to anticipate catastrophe (Beck 1999).

A risk society is one that emerges from modernization, and it is structured through specialized knowledge and a mathematical moral of expert thinking (Beck 2006). Risk is policy-oriented, and it can be positioned as a risk management. Risk management implies that it is the product of the probability of occurrence multiplied by the intensity and scope of potential harm.

Risks are products of society's perception of the dangers surrounding it. Risks are selected by various groups for attention. As a collective construct, risks can be selected and considered or deemed acceptable and disregarded, depending on how society perceives and problematizes it (Douglas and Wildavsky 1983). Risks arise from uncertainty and have a cultural framework. Risks are presented as expert knowledge, but each person's perception will depend on how people build it socially. In the contemporary world, the framework of a global culture tends to reinforce risk-management culture. An essential feature of risk societies is that the consequences of their management will always be ambivalent (Beck 2006).

In this context of risk society, the digital world reproduces a permanent risk, constantly creating new risk modalities that shake the institutional foundations of governments and firms and the ability of organizations to anticipate global catastrophes. Risk perceptions have three characteristics (Beck 2006) that can be applied to the digital world. First, they are delocalized. This means that the causes and consequences of using digital tools are not limited to a geographical location. For example, more and more governments are operating their data in clouds that are owned by private companies and governed by contracts that organize a virtual government infrastructure. Second, the digital world favors incalculable risk. Attempts at securitization and control are not enough at the stage of knowledge for the contemporary world. For example, companies are looking for different forms of insurance to cover the costs of cyberattacks (Satariano and Perlroth 2019). Contemporary societies virtualize risk. Finally, in a world of virtual risks, the consequences of disasters are not controllable and compensable. The interdependence of the digital world maximizes risk because of low control and anticipation.

Think tanks try to map the risks from the digital world and their application in public policy and services. Digital transformation means that organizations take more risks and that they need to adapt in a context of increasing turbulence. Deloitte has mapped risks of the digital world by associating the characteristics of a risk society.

- The use of AI by organizations operates as black boxes for decision making. Artificial intelligence produces decisions that cannot be explained, making it difficult to detect inappropriate decisions. These expose organizations to vulnerabilities such as biases, unsuitable techniques, and incorrect decisions made by algorithms. Organizations should consider ethical elements, fairness, and safety in how algorithms are used.
- Automating service processes and public policy can produce unintended consequences such as obsolescence of controls, complexity of operations, and the possibility of cascading errors.
- Adopting technology has unintended consequences. Although this approach seeks increased efficiency in policies and services, it adds cybersecurity challenges. Cyberattacks are facilitated by the scale of digital tools' use.
- Using digital tools creates conditions for manipulating information at a large scale. Social media influence public opinion with the widespread use of digital tools based on false perceptions. Misinformation magnifies society's risks and creates errors in decision making.
- Managing data becomes riskier when organizations begin to collect, store, and monetize data. Organizations can use data for information warfare and reduce their ability to create public value.
- Using new technologies makes organizations more dependent on third parties.

- Invoking regulatory changes leads to a focus on available technologies rather than emerging ones, thereby requiring more flexible and less risk-mitigating regulatory instruments.
- Aiming for digital transformation poses new cultural risks when organizational goals and employee values and behaviors are misaligned.
- Advancing digital technologies comes at a price. Organizations must deal with expectations of ethical obligations, social responsibilities, and new organizational values (Deloitte 2019).

Additionally, organizations are subject to misuse and digital illiteracy by users, requiring policies that combat the digital divide. Finally, digitization of public services and policy can present design problems, reinforcing government failures. These risks are neither exhaustive nor definitive, but they are inherent to digital transformation and point to a constitutive problem of digital governance. Adopting digital tools to improve governance is important for improving the efficiency and effectiveness of public policies and services. However, the experimental and disruptive character of digital innovations can lead to problems of undersupply of services and ineffectiveness in policy, generating crises of trust and state illegitimacy.

The problem with trust begins when institutions mobilize their ethical and normative resources, especially for political institutions. Trust in institutions is the process of evaluation and judgment as carried out by public opinion. The inexistence of trust in institutions derives from the limited information available to individuals regarding politics. Distrust is a symptom of information asymmetry and the sparse cognitive resources available to public opinion (Hardin 1999). On the other hand, distrust results from the incoherence of institutions relative to their normative resources (Offe 1999). Trust in institutions presumes that society is aware of their basic norms and their permanent functions, in view of the normative values and purposes that surround them (Inglehart and Welzel 2005).

Distrust jeopardizes the government's capacity to coordinate society and establish social cooperation (Newton 1999). Furthermore, distrust toward institutions does not allow for a consolidation of a democratic political culture grounded on these institutions' ability to build their normative justification (Levi 1998; Dalton 1999). The risks of organizations adopting digital tools to improve their governance imply that institutions take risks and that they will depend on more governance for digital tools. This governance takes place in a context of turbulent organizations that use adaptation strategies to test their resilience and transform their basic structures and normative functions.

The greatest risk for digital for governance is to foster a crisis of distrust in institutions and belief in illegitimacy of the state, companies, and organizations. Digital governance can improve the context of policy implementation and service delivery.

References

Agranoff, R., & McGuire, M. (2004). *Collaborative Public Management. New Strategies for Local Governments*. Washington: Georgetown University Press.

Ahn, M., & Bretschneider, S. (2011). Politics of E-Government: E-Government and the Political Control of Bureaucracy. *Public Administration Review, 71*(3), 414–424. https://doi.org/10.1111/j.1540-6210.2011.02225.x.

Alkhatib, A., & Bernstein, M. (2019, May 4–9). *Street-Level Algorithms: A Theory at the Gaps Between Policy and Decisions*. Proceedings of the 2019 CHI Conference on Human Factors in Computing Society Systems Proceedings, Glasgow, Scotland. New York: ACM. https://doi.org/10.1145/3290605.3300760.

Allcott, H., & Gentzkow, M. (2017). Social Media and Fake News in the 2016 Election. *Journal of Economic Perspectives, 31*(2), 211–236. https://doi.org/10.1257/jep.31.2.211.

Almeida, V., Filgueiras, F., & Gaetani, F. (2019). Principles and Elements of Governance of Digital Public Services. *IEEE Internet Computing, 23*(6), 48–53.

Andrews, L. (2018). Public Administration, Public Leadership and the Construction of Public Value in the Age of the Algorithm and 'Big Data'. *Public Administration, 97*(2), 296–310. https://doi.org/10.1111/padm.12534.

Ansell, C., & Trondal, J. (2018). Governing Turbulence: An Organizational-Institutional Agenda. *Perspectives on Public Management and Governance, 1*(1), 43–57. https://doi.org/10.1093/ppmgov/gvx013.

Attard, J., Orlandi, F., Scerri, S., & Auer, S. (2015). A Systematic Review of Open Government Data Initiatives. *Government Information Quarterly, 32*(4), 399–418. https://doi.org/10.1016/j.giq.2015.07.006.

Balutis, A. P. (2001). E-Government 2001. Part I: Understanding the Challenge and Evolving Strategies. *The Public Manager, 30*(1), 33–37.

Bannister, F., & Connolly, R. (2011). Trust and Transformational Government: A Proposed Framework for Research. *Government and Information Quarterly, 28*(2), 137–147. https://doi.org/10.1016/j.giq.2010.06.010.

Bartje, J. (2016). The Top 10 Application Areas—Based on Real IoT Projects. *IoT Analytics*. Retrieved from https://iot-analytics.com/top-10-iot-project-application-areas-q3-2016/.

Beck, U. (1999). *World Risk Society*. Cambridge: Polity Press.

Beck, U. (2006). Living in the World Risk Society. A Hobhouse Memorial Public Lecture Given on Wednesday 15 February 2006 at the London School of Economics. *Economy and Society, 35*(3), 329–345. https://doi.org/10.1080/03085140600844902.

Berghaus, S., & Back, A. (2016). *The Fuzzy Front-End of Digital Transformation: Three Perspectives on the Formulation of Organizational Change Strategies*. In: BLED 2016 Proceedings, Bled E-Conference, Bled, Slovenia, 129–144. Available at: https://aisel.aisnet.org/bled2016/40/.

Bertot, J., Estevez, E., & Janowski, T. (2016). Universal and Contextualized Public Services: Digital Public Service Innovation Framework. *Government Information Quarterly, 33*(2), 211–222. https://doi.org/10.1016/j.giq.2016.05.004.

Boulos, K., & Geraghty, E. (2020). Geographical Tracking and Mapping of Coronavirus Disease COVID-19/Severe Acute Respiratory Syndrome Coronavirus 2 (SARS-CoV-2) Epidemic and Associated Events Around the World: How 21st Century GIS Technologies are Supporting the Global Fight Against Outbreaks and Epidemics. *International Journal of Health Geographics, 19*(8). https://doi.org/10.1186/s12942-020-00202-8.

Brewer, G. A., Neubauer, B. J., & Geiselhart, K. (2006). Designing and Implementing E-Government Systems: Critical Implications for Public Administration and Democracy. *Administration & Society, 38*(4), 472–499. https://doi.org/10.1177/0095399706290638.

Chen, Y. C. (2017). *Managing Digital Governance. Issues, Challenges and Solutions*. New York: Routledge.

Chen, Y. F., & Hsieh, T. C. (2014). Big Data for Digital Government: Opportunities, Challenges, and Strategies. *International Journal of Public Administration in the Digital Age, 1*(1), 1–14. https://doi.org/10.4018/ijpada.2014010101.

Chen, D., & Zhao, H. (2012). *Data Security and Privacy Protection Issues in Cloud Computing*. Proceedings of the 2012 International Conference on Computer Science and Electronics Engineering, Hangzhou, IEEE. https://doi.org/10.1109/ICCSEE.2012.193.

Chun, S. A., Shulman, S., Sandoval, R., & Hovy, E. (2010). Government 2.0: Making Connections Between Citizens, Data and Government. *Information Polity, 15*(1/2), 1–9. https://doi.org/10.3233/IP-2010-0205.

Ciborra, C. (2004). *Digital Technologies and the Duality of Risk*. CARR Discussion Papers (DP 27). Centre for Analysis of Risk and Regulation, London School of Economics and Political Science, London. Retrieved from http://eprints.lse.ac.uk/36069/.

Cohen, M. D., March, J. G., & Olsen, J. P. (1972). A Garbage Can Model of Organizational Choice. *Administrative Science Quarterly, 17*(1), 1–25. https://doi.org/10.2307/2392088.

Creemers, R. (2018). *China's Social Credit System: An Evolving Practice of Control.* SSRN. Retrieved from https://doi.org/10.2139/ssrn.3175792.

Dahlberg, L. (2007). Civic Identity and Net Activism. The Frame of Radical Democracy. In L. Dahlberg & E. Siapera (Eds.), *Radical Democracy and the Internet* (pp. 55–72). New York: Palgrave.

Dalton, R. (1999). Political Support in Advanced Industrial Democracies. In P. Norris (Ed.), *Critical Citizens. Global Support for Democratic Government.* Cambridge: Oxford University Press.

Danaher, J. (2016). The Threat of Algocracy: Reality, Resistance and Accommodation. *Philosophy & Technology, 29*(3), 245–268. https://doi.org/10.1007/s13347-015-0211-1.

Danaher, J., Hogan, M. J., Noone, C., Kennedy, R., Behan, A., De Paor, A., Felzman, H., Haklay, M., Khoo, S. M., Morison, J., Murphy, M. H., O'Brolchain, N., Schafer, B., & Shankar, K. (2017). Algorithmic Governance: Developing a Research Agenda Through the Power of Collective Intelligence. *Big Data & Society, 4*(2), 1–21. https://doi.org/10.1177/2053951717726554.

Dawes, S. S. (2009). Governance in the Digital Age: A Research and Action Framework for an Uncertain Future. *Government Information Quarterly, 26*(2), 257–264. https://doi.org/10.1016/j.giq.2008.12.003.

Deloitte. (2019). *Future of Risk in the Digital Era.* Transformative Change. Disruptive Risk. New York: Deloitte. Retrieved from https://www2.deloitte.com/us/en/pages/advisory/articles/risk-in-the-digital-era.html.

Domingos, P. (2015). *The Master Algorithm. How the Quest for Ultimate Machine Learning will Remake Our World.* New York: Basic Books.

Douglas, M., & Wildavsky, A. (1983). *Risk and Culture. An Essay on the Selection of Technical and Environmental Dangers.* Berkeley: University of California Press.

Duguit, L. (1923). The Concept of Public Service. *Yale Law Journal, 32*(5), 425–435. https://doi.org/10.2307/788739.

Dunleavy, P. (2016). Big Data and Policy Learning. In G. Stoker & M. Evans (Eds.), *Evidence-Based Policy Making in the Social Sciences. Methods that Matter.* Chicago: Policy Press.

Dunleavy, P., & Margetts, H. (2013). The Second Wave of Digital-Era Governance: A Quasi-Paradigm for Government on the Web. *Philosophical Transactions of the Real Society, 371*(1987), 1–17. https://doi.org/10.1098/rsta.2012.0382.

Dunleavy, P., Margetts, H., Bastow, S., & Tinkler, J. (2006). New Public Management Is Dead—Long Live Digital-Era Governance. *Journal of Public Administration Research and Theory, 16*(3), 467–494. https://doi.org/10.1093/jopart/mui057.

Dye, T. (1972). *Understanding Public Policy.* Englewood Cliffs, NJ: Prentice-Hall.

Earley, S. (2014). The Digital Transformation: Staying Competitive. *IT Professional, 16*(2), 58–60. https://doi.org/10.1109/MITP.2014.24.

Eggers, W. D., Schatsky, D., & Viechnicki, P. (2017). *AI-Augmented Government.* Using Cognitive Technologies to Redesign Public Sector Work. Deloitte Center for Government Insights. Retrieved from https://www2.deloitte.com/insights/us/en/focus/cognitive-technologies/artificial-intelligence-government.html.

Etzioni, A. (2010). Is Transparency the Best Disinfectant? *The Journal of Political Philosophy, 18*(3), 389–404. https://doi.org/10.1111/j.1467-9760.2010.00366.x.

Ferro, E., Helbig, N. C., & Gil-Garcia, J. R. (2011). The Role of IT Literacy in Defining Digital Divide Policy Needs. *Government Information Quarterly, 28*(1), 3–10. https://doi.org/10.1016/j.giq.2010.05.007.

Filgueiras, F., Cireno, F., & Palotti, P. (2019). Digital Transformation and Public Service Delivery in Brazil. *Latin American Policy, 10*(2), 195–219. https://doi.org/10.1111/lamp.12169.

Fountain, J. E. (2004). *Building the Virtual State. Information Technology and Institutional Change.* Washington, DC: Brookings Institution Press.

Gasser, U., & Almeida, V. (2017). A Layered Model for AI Governance. *IEEE Internet Computing, 21*(6), 58–62. https://doi.org/10.1109/mic.2017.4180835.

Gillespie, T. (2010). The Politics of 'Platforms'. *New Media & Society, 12*(3), 347–364. https://doi.org/10.1177/1461444809342738.

Gillingham, P. (2015). Predictive Risk Modelling to Prevent Child Maltreatment and Other Adverse Outcomes for Service Users: Inside the 'Black Box' of Machine Learning. *The British Journal of Social Work, 46*(4), 1044–1058. https://doi.org/10.1093/bjsw/bcv031.

Greengard, S. (2015). *The Internet of Things.* Cambridge, MA: MIT Press.

Hansen, A. M., Kræmmegaard, P., & Mathiasen, L. (2011). Rapid Adaptation in Digital Transformation: A Participatory Process for Engaging IS and Business Leaders. *MIS Quarterly Executive, 10*(4), 175–185. Retrieved from https://aisel.aisnet.org/misqe/vol10/iss4/5.

Hardin, R. (1999). Do We Want Trust in Government? In M. Warren (Ed.), *Democracy and Trust.* Cambridge: Cambridge University Press.

Heald, D. (2006). Transparency as an Instrumental Value. In C. Hood & D. Heald (Eds.), *Transparency: The Key of Better Governance.* Oxford: Oxford University Press.

Helmond, A. (2015). The Platformization of the Web: Making Web Data Platform Ready. *Social Media + Society, 1*(2), 1–11. https://doi.org/10.1177/2056305115603080.

Hood, C., & Margetts, H. (2007). *The Tools of Government in the Digital Age.* London: Palgrave Macmillan.

Howlett, M. (2005). What Is a Policy Instrument? Tools, Mixes, and Implementation Styles. In P. Eliadis, M. G. Hill, & M. Howlett (Eds.), *Designing Government. From Instruments to Governance* (pp. 31–50). Montreal: McGill-Queens' University Press.

Howlett, M. (2014). Conceptualizing Public Policy. In I. Engeli & C. R. Allison (Eds.), *Comparative Policy Studies* (pp. 17–33). London: Palgrave Macmillan. https://doi.org/10.1057/9781137314154_2.

Howlett, M. (2019). Procedural Policy Tools and the Temporal Dimensions of Policy Design: Resilience, Robustness and the Sequencing of Policy Mixes. *International Review of Public Policy, 1*(1), 27–45. https://doi.org/10.4000/IRPP.310.

Inglehart, R., & Welzel, C. (2005). *Modernization, Cultural Change, and Democracy. The Human Development Sequence*. New York: Cambridge University Press.

Janssen, D., & Kies, R. (2005). Online Forums and Deliberative Democracy. *Acta Politica, 40*(3), 384–392. https://doi.org/10.1057/palgrave.ap.5500115.

Janssen, M., & Kuk, G. (2016). The Challenges and Limits of Big Data Algorithms in Technocratic Governance. *Government Information Quarterly, 33*(3), 371–377. https://doi.org/10.1016/j.giq.2016.08.011.

Janssen, M., Charalabidis, Y., & Zuiderwijk, A. (2012). Benefits, Adoption Barriers and Myths of Open Data and Open Government. *Information Systems Management, 29*(4), 258–268. https://doi.org/10.1080/10580530.2012.716740.

Janssen, M., Rana, N. P., Slade, E., & Dwivedi, Y. K. (2018). Trustworthiness of Digital Government Services: Deriving a Comprehensive Theory through Interpretive Structural Modelling. *Public Management Review, 20*(5), 647–671. https://doi.org/10.1080/14719037.2017.1305689.

Jordan, T. (2007). Online Direct Action: Hacktivism and Radical Democracy. In L. Dahlberg & E. Siapera (Eds.), *Radical Democracy and the Internet* (pp. 73–88). New York: Palgrave.

Just, N., & Latzer, M. (2017). Governance by Algorithms: Reality Construction by Algorithmic Selection on the Internet. *Media Culture & Society, 39*(2), 238–258. https://doi.org/10.1177/0163443716643157.

Kang, J. S., Kuznetsova, P., Luca, M., & Choi, Y. (2013). Where Not to Eat? Improving Public Policy by Predicting Hygiene Inspections Using Online Reviews. Proceedings of the 2013 Conference on Empirical Methods in Natural Language Processing. Seattle: Association for Computational Linguistics (ACL), pp. 1443–1448. Retrieved from https://www.aclweb.org/anthology/D13-1150/.

Kitchin, R. (2014). The Real-Time City? Big Data and Smart Urbanism. *GeoJournal, 79*(1), 1–14. https://doi.org/10.1007/s10708-013-9516-8.

König, P. D. (2019). Dissecting the Algorithmic Leviathan: On the Social-Political Autonomy of Algorithmic Governance. *Philosophy & Technology, 32*(4). https://doi.org/10.1007/s13347-019-00363-w.

Kosters, M., & Van der Heijden, J. (2015). From Mechanism to Virtue: Evaluating Nudge Theory. *Evaluation, 21*(3), 276–291. https://doi.org/10.1177/1356389015590218.

Legner, C., Eymann, T., Hess, T., Matt, C., Böhmann, T., Drews, P., Mädche, A., Urbach, N., & Ahlemann, F. (2017). Digitalization: Opportunity and Challenge for the Business and Information Systems Engineering Community. *Business & Information Systems Engineering, 59*(4), 301–308. https://doi.org/10.1007/s12599-017-0484-2.

Levi, M. (1998). A State of Trust. In V. Braithwaite & M. Levi (Eds.), *Trust and Governance*. New York: Russell Sage Foundation.

Linders, D. (2012). From E-Government to We-Government: Defining a Typology for Citizen Coproduction in the Age of Social Media. *Government Information Quarterly, 29*(4), 446–454. https://doi.org/10.1016/j.giq.2012.06.003.

Lindstedt, C., & Naurin, D. (2010). Transparency Is Not Enough: Making Transparency Effective in Reducing Corruption. *International Political Science Review., 31*(3), 301–322. https://doi.org/10.1177/0192512110377602.

Lipsky, M. (2010). *Street-Level Bureaucracy. Dilemmas of the Individual in Public Service*. New York: Russell-Sage Foundation.

Luna-Reyes, L. F., & Gil-Garcia, J. R. (2014). Digital Government Transformation and Internet Portals: The Co-Evolution of Technology, Organizations, and Institutions. *Government Information Quarterly, 31*(4), 545–555. https://doi.org/10.1016/j.giq.2014.08.001.

Luna-Reyes, L. F., Gil-Garcia, J. R., & Cruz, C. B. (2007). Collaborative Digital Government in Mexico: Some Lessons from Federal Web-Based Interorganizational Information Integration Initiatives. *Government Information Quarterly, 24*(4), 808–826. https://doi.org/10.1016/j.giq.2007.04.003.

Madakam, S., Ramaswamy, R., & Tripathi, S. (2015). Internet of Things (IoT): A Literature Review. *Journal of Computer and Communications, 3*(5), 164–173. https://doi.org/10.4236/jcc.2015.35021.

Maedche, A. (2016). Interview with Michael Nilles on "What Makes Leaders Successful in the Age of the Digital Transformation?". *Business & Information Systems Engineering, 58*(4), 287–289. https://doi.org/10.1007/s12599-016-0437-1.

March, J., & Olsen, J. P. (1989). *Rediscovering Institutions. The Organizational Basis of Politics*. New York: Free Press.

Margetts, H. (2013). Data, Data Everywhere: Open Data Versus Big Data in the Quest for Transparency. In N. Bowles, J. Hamilton, & D. Levy (Eds.),

Transparency in Politics and the Media. Accountability and Open Government. London: IB Tauris.
McDermott, P. (2010). Building Open Government. *Government Information Quarterly, 27*(4), 401–413. https://doi.org/10.1016/j.giq.2010.07.002.
Meijer, A. J. (2012). Co-Production in an Information Age: Individual and Community Engagement Supported by New Media. *Voluntas: International Journal of Voluntary and Nonprofit Organizations, 23*(4), 1156–1172. https://doi.org/10.1007/s11266-012-9311-z.
Mergel, I., Edelman, N., & Haug, N. (2019). Defining Digital Transformation: Results from Experts Interviews. *Government Information Quarterly, 36*(4). https://doi.org/10.1016/j.giq.2019.06.002.
Mikhaylov, S. J., Esteve, M., & Campion, A. (2018). Artificial Intelligence for Public Sector: Opportunities and Challenges of Cross-Sector Collaboration. *Philosophical Transactions of the Royal Society, 376*(2128), 1–26. https://doi.org/10.1098/rsta.2017.0357.
Milakovich, M. E. (2012). *Digital Governance: New Technologies for Improving Public Service and Participation.* New York: Routledge. https://doi.org/10.4324/9780203815991.
Moynihan, D. P. (2006). Ambiguity in Policy Lessons: The Agencification Experience. *Public Administration, 84*(4), 1029–1050. https://doi.org/10.1111/j.1467-9299.2006.00625.x.
Napoli, P. M. (2014). Automated Media: An Institutional Theory Perspective on Algorithmic Media Production and Consumption. *Communication Theory, 24*(3), 340–360. https://doi.org/10.1111/comt.12039.
Newton, K. (1999). The Impact of Social Trust on Political Support. In P. Norris (Ed.), *Critical Citizens. Global Support for Democratic Government.* Oxford: Oxford University Press.
Nord, J. H., Koohang, A., & Paliszkewicz, J. (2019). The Internet of Things: Review and Theoretical Framework. *Expert Systems with Applications, 133*(1), 97–108. https://doi.org/10.1016/j.eswa.2019.05.014.
O'Reilly, T. (2010). Government as Platform. In D. Lahtrop & L. Ruma (Eds.), *Open Government: Collaboration, Transparency, and Participation in Practice.* Sebastopol: O'Reilly Media.
Offe, C. (1999). How Can We Trust Our Fellow Citizens? In M. Warren (Ed.), *Democracy and Trust.* Cambridge: Cambridge University Press.
Olsen, J. P. (2007). *Europe in Search of Political Order.* Oxford: Oxford University Press.
Olsen, J. P. (2010). *Governing through Institutional Building.* Oxford: Oxford University Press.
Perl, A., Howlett, M., & Ramesh, M. (2018). Policy-Making and Truthiness: Can Existing Policy Models Cope with Politicized Evidence and Willful Ignorance in a "Post-Fact" World? *Policy Sciences, 51*(4), 581–600. https://doi.org/10.1007/s11077-018-9334-4.

Peters, B. G. (2005). Policy Instruments and Policy Capacity. In M. Painter & J. Pierre (Eds.), *Challenges to State Policy Capacity*. London: Palgrave Macmillan.

Salamon, L. (2001). The New Governance and the Tolls of Public Action: An Introduction. *Fordham Urban Law Journal, 28*(5), 1611–1674. Retrieved from https://ir.lawnet.fordham.edu/ulj/vol28/iss5/4.

Samuel, A. (1959). Some Studies in Machine Learning Using the Game of Checkers. *IBM Journal of Research and Development, 44*(1.2), 210–229. https://doi.org/10.1147/rd.33.0210.

Sandoval-Almazán, R., Luna-Reyes, L. F., Luna-Reyes, D., Gil-Garcia, J. R., Puron-Cid, G., & Picazo-Vela, S. (2017). *Building Digital Government Strategies. Principles and Practices*. Cham: Springer Nature.

Satariano, A., & Perlroth, N. (2019, April 15). Big Companies Thought Insurance Covered a Cyberattack. They May Be Wrong. *The New York Times*. Retrieved from https://www.nytimes.com/2019/04/15/technology/cyberinsurance-notpetya-attack.html.

Scheerder, A., van Deursen, A., & van Dijk, J. (2017). Determinants of Internet Skills, Uses and Outcomes: A Systematic Review of the Second and Third-Level Digital Divide. *Telematics & Informatics, 34*(8), 1607–1624. https://doi.org/10.1016/j.tele.2017.07.007.

Seo, D. (2017). Digital Business Convergence and Emerging Contested Fields: A Conceptual Framework. *Journal of the Association for Information Systems, 18*(10), 687–702. https://doi.org/10.17705/1jais.00471.

Servick, K. (2020). Cellphone Tracking Could Help Stem the Spread of Coronavirus. Is Privacy the Price? *Science*. https://doi.org/10.1126/science.abb8296.

Shackelford, S., & Myers, S. (2018). Block-by-Block: Leveraging the Power of Blockchain Technology to Build Trust and Promote Cyber Peace, 19 Yale J.L. & Tech (2018). Retrieved from https://digitalcommons.law.yale.edu/yjolt/vol19/iss1/7.

Simon, H. (1969). *The Science of Artificial*. Cambridge: MIT Press.

Stoker, G. (2019). Embracing Complexity: A Framework for Exploring Governance Resources. *Journal of Chinese Governance., 4*(2), 91–107. https://doi.org/10.1080/23812346.2019.1587859.

Tammpuu, P., & Masso, A. (2018). "Welcome to the Virtual State": Estonia E-Residency and the Digitalized State as a Commodity. *European Journal of Cultural Studies, 21*(5), 543–560. https://doi.org/10.1177/1367549417751148.

Tassabehji, R., Hackney, R., & Popovic, A. (2016). Emergent Digital Era Governance: Enacting the Role of the 'Institutional Entrepreneur' in Transformational Change. *Government Information Quarterly, 33*(2), 223–236. https://doi.org/10.1016/j.giq.2016.04.003.

Veale, M., & Brass, I. (2019). Administration by Algorithm? Public Management Meets Public Sector Machine Learning. In K. Yeung & M. Lodge (Eds.), *Algorithm Regulation*. Oxford: Oxford University Press.

Vial, G. (2019). Understanding Digital Transformation: A Review and a Research Agenda. *Journal of Strategic Information Systems, 28*(2), 118–144. https://doi.org/10.1016/j.jsis.2019.01.003.

Weber, R. H. (2013). Internet of Things—Governance Quo Vadis? *Computer Law & Security Review, 29*(4), 341–347. https://doi.org/10.1016/j.clsr.2013.05.010.

Weerakkody, V., Omar, A., El-Haddadeh, R., & Al-Busaidy, M. (2016). Digitally-Enabled Service Transformation in the Public Sector: The Lure of Institutional Pressure and Strategic Response Towards Change. *Government Information Quarterly, 33*(4), 658–668. https://doi.org/10.1016/j.giq.2016.06.006.

Westerman, G. (2016). Why Digital Transformation Needs a Heart. *MIT Sloan Management Review, 58*(1), 10–13.

Wildavsky, A. B. (1988). *Searching for Safety*. New Jersey: Transaction Publishers.

Williamson, B. (2014). Knowing Public Services: Cross-Sector Intermediaries and Algorithmic Governance in Public Sector Reform. *Public Policy & Administration, 29*(4), 292–312. https://doi.org/10.1177/0952076714529139.

Wu, D., Si, S., Wu, S., & Wang, R. (2018). Dynamic Trust Relationships Aware Data Privacy Protection in Mobile Crowd-Sensing. *IEEE Internet of Things Journal, 5*(4), 2958–2970. https://doi.org/10.1109/JIOT.2017.2768073.

CHAPTER 4

Governance for Digital Technologies

Abstract Digital technologies are important tools for supporting economic development and collective action and solving relevant public problems. On the other hand, they have many associated risks, requiring governance mechanisms to protect vulnerable groups of society and to enhance public well-being. This chapter focuses on the design of institutions essential to governing the digital world. It also presents theoretical concepts that might be part of the construction of a digital governance framework.

Keywords Institution • Grammar • Polycentric governance • Cybersecurity • Multistakeholder • Decision-making process • Transparency • Accountability • Coordination • Self-governance • Adaptive capacity

The previous chapter discussed how digital tools can help build better governance processes. This chapter focuses on governance mechanisms and processes for digital tools and resources that constitute the digital world's core. A famous aphorism mistakenly attributed to Marshall McLuhan says: "We become what we behold. We shape our tools and then our tools shape us." The quote was actually written by Father John Culkin (Culkin 1967). Tools such as Google, Waze, Facebook, LinkedIn, Twitter, Alexa, WeChat, YouTube, Spotify, Uber, and WhatsApp are powerful examples that illustrate how digital resources shape society and create new patterns of behavior around the world. The digital world is still a confusing scenario with many unanswered questions—especially related to the role of governments, society, and big tech companies in the creation of rules and models for oversight, transparency, accountability, and management of global digital platforms.

Current and future digital tools must be governed so that their use and ownership by governments, firms, and society can be applied to solve some of society's biggest challenges. The digital world requires governance in a way that promotes and protects human rights, and preserves ethical values and society's well-being (IEEE 2019). While governance has the potential to promote digital technologies to improve human organizations, governance also has to find ways to address the complexity and ubiquity of digital technologies. One of the challenges is to find out which institutions need to be designed to govern the digital world to achieve the desired results.

The design of institutions to govern the digital world faces a context of increasing adaptation. Adaptation requires an institutional dynamic to be able to respond to specific challenges associated with the growth of the digital world. The design of institutions should consider the inherent uncertainties of the digital world, caused by a number of factors: fast pace of technology evolution, quasi-permanent turbulence in institutional environment, conditions beyond historical experience, and a diversity of impacts on different groups of society. In short, this chapter concentrates on principles, models, and norms essential to building the complex institutional design to govern the digital world.

Designing Institutions

The results of the use of digital tools for governing are twofold. Digital tools provide benefits for public policy and services and promote a perspective of smart governance. They are important tools for supporting economic development and collective action and solving relevant public problems (Bolívar and Meijer 2015). However, digital tools have many associated risks, requiring governance mechanisms to protect vulnerable groups of society and to enhance public well-being.

Defining these governance mechanisms requires thinking about the type of institutional design to be applied to the digital world. Definitions, in this case, are essential for the proposed solutions to achieve what they want. Design "is a creation of actionable form to promote valued outcomes in a particular context" (Bobrow and Dryzek 1987; Alexander 1964). Designing institutions requires, from this perspective, outlining the specific results to be achieved, determining which values are embedded in the practices to be fostered through the institutions, and discerning the rules and norms needed to achieve the intended result (Goodin 1996).

Institutions are designed to delimit individuals' behavior. They can play an essential role in society, the economy, and politics, because they promote incentives and constraints for individual action and equilibria solutions that enable collective action and solidarity (March and Olsen 1984). Human organization implies a political order capable of carrying out the highest human values (Goodin 1996). The institutions are designed in the context of a transformation process to define a grammar to the various actors involved in different public problems. Institutional grammars are regularities of human action in situations structured by rules, norms, and shared strategies. These rules, norms, and strategies are constituted and reconstituted by frequent and repetitive actions that are incorporated in the various situations of human life. Syntax of the grammar identifies institutional components and enables the analysis of cooperation in situations of action dilemmas (Crawford and Ostrom 1995). In this sense, constituting an institutional grammar does not just mean creating legislation or standards. It means how many of these rules fit into the actors' daily practice.

In defining a grammar, institutions are designed to promote the results of implicit or explicit efforts by a set of individuals to achieve order and predictability within defined situations by (1) creating positions; (2) stating how participants enter or leave positions; (3) explaining which actions participants are required, permitted, or forbidden to take; (4) specifying which outcome participants are required, permitted, or forbidden to affect; and (5) stipulating the consequences of rule violation, which in most cases we expect to be associated to a specific sanction (Crawford and Ostrom 1995). Institutions are strategies, norms, and rules that indicate permitted, required, or forbidden conduct within specified temporal, spatial, and procedural boundaries. Institutions are essential to the governance of individual and collective behavior in society and political and economic order (Siddiki et al. 2019).

Institutional design establishes and reflects society's expectations of the values to be achieved and who can do what, where, and how to solve collective action problems and governance challenges (Ostrom 2005). The design of the institutions is not static. Institutions are in a constant dynamic of transformation and change, adapting to new contexts or promoting disruption by introducing new grammar (Ostrom 2005). Transformation can occur by accident, evolution, or intentionally (Goodin 1996). These three forms are not necessarily exhaustive or exclusive. The dynamics of change can involve these three elements at the same time, requiring from the various actors' exhaustive designs and resilience to adapt new solutions. The big challenge for governance of the digital world is to create a new order. In other words, the digital world needs to be governed in a new grammar, with attributes to adapt and produce an order for a complex and new world.

The values given to the tragedy of the digital world require the protection of human rights (Donahoe and Metzger 2019), greater effectiveness of services and public policies, well-being, and greater control over the possibility of misuse of technologies (IEEE 2019). Designing institutions for governance in the digital world requires designing new institutions and changing the existing ones, within a framework motivated by complex solutions to a complex world.

Institutional grammars for the digital world are fragmented, broken down into several languages to solve specific challenges. The governance of the digital world is a complex framework, in constant adaptation, with different rules, norms, space, and time. Joseph Nye describes a big picture of norms, institutions, and procedures that are part of the regime complex for managing global cyberactivities (Nye 2014). There are many international institutions in the cyber governance map. Some are big, others are small, some are formal organizations, and other are informal. Some are multilateral organizations—such as the Group of Twenty (G20), UN-ITU, and OECD—while others are multistakeholder or non-governmental institutions, such as ICANN and the Electronic Frontier Foundation (EFF). This fragmentation of digital governance into various issues creates difficulty in establishing a common institutional grammar.

The digital world is changing at a speed that makes it difficult to establish a robust institutional grammar. Digital technologies are creating a scenario in which organizational turbulence becomes the new normal (Ansell et al. 2017). Technologies test the resilience of institutions in the political, economic, and social order, so that solutions are always motivated by specific issues. An institutional grammar should promote, first and foremost, a holistic view of the governance problem to be faced, to promote balanced solutions and a common language for the various actors. Defining an institutional grammar for the digital world is critical to specifying actors' roles, the rules and norms, and strategies that enable forms of collective action to overcome the tragedy of the digital world.

Governance for the digital world lacks this holistic view, fragmenting into various issues or components without a common grammar. Internet governance is a key issue for the security of societies, for economic relationships, and for availability of digital resources to the various actors. Internet governance takes care of the distribution of critical resources, as well as the communication mechanisms that feed new economic relations (DeNardis 2014). Digital platform governance is a response to the growing platformization of societies (Helmond 2015), aiming at regulating communication and content on social media (Gillespie 2010). Artificial intelligence governance reflects concern about discrimination, bias, and technological redlining problems, because of the use of AI and autonomous decision systems (Gasser and Almeida 2017). Similarly, IoT governance seeks to govern the situations in which devices obtain data and decide (Almeida et al. 2015).

Cybersecurity issues are global, requiring increased regulation in the context of complex regimes involving global trade, geopolitics, and interests to create a context of interdependence for cyber activities (Nye 2014). The Global Commission on the Stability of Cyberspace developed norms and policies to enhance international security and stability and orient responsible state and non-state behavior in cyberspace (Klimburg and Almeida 2019). The final report outlines eight norms and a set of principles for behavior in cyberspace. It includes norms such as the one demanding that state and non-state actors must not pursue, support, or allow cyber operations intended to disrupt the technical infrastructure essential to elections, referenda, or plebiscites. Another norm regarding the protection of the public core of the Internet urges state and non-state actors to avoid attacks that would intentionally and substantially damage the general availability or integrity of the "public core" of the Internet (GCSC 2019). Finally, data governance is a response to the growing datafication of society, by seeking common standards and norms for data storage, organizational structures, policies and standards, data communication and management, and quality measures that regulate privacy and democracy issues (Al-Badi et al. 2019).

These are the basic issues of digital governance. Other issues may emerge as new technologies (such as facial recognition or autonomous vehicles) enter the global market. It is expected that the introduction of autonomous vehicles will create a whole new set of problems, demanding new institutional solutions for the control and well-being of society (Millard-Ball 2016). The digital world requires redundant solutions in different jurisdictions. These redundant solutions become a possibility to govern a world that is virtual, but with concrete consequences for society and people. In the context of governance, institutions and policies should be effective to establish an order for the digital world—from political, economic, and social perspectives. The rapid evolution of digital technologies provides a context of constant reform and permanent adaptation, given the conditions of digital transformation.

All these issues gradually convert into a holistic solution. Theoretically, it is necessary to build a common institutional grammar for solving the tragedy of the digital world. Figure 4.1 illustrates a digital governance framework that must be theorized to promote the institutional grammar, with different elements of a common morphology to the various governance challenges for the digital world. This governance arrangement is non-territorial, involving multiple actors seeking solutions that can be local, regional, or global, in a period that requires constant adaptation to build resilience. This governance arrangement does not have a defined and centralized authority and should not be governed by an "invisible hand," considering all associated risks.

The big challenge is to design institutions to govern the digital world in a common grammar. This involves multiple actors with distinct interests in the process of governing the digital world. These multiple actors, such as government bureaucracies, economic actors, military, intelligence agencies, civil society organizations, and the press participate in making decisions. The challenge is to design institutions that steer these multiple actors' behavior, with different points of authority, to promote public interest and cooperative conditions.

The characteristic of digital governance and the way it takes on institutional design point to its polycentric nature. Polycentric governance is a grammar characterized by multiple decision centers and multiple issues. First, the decision-making process occurs between various levels and functions of government operating in a shared decision system (Ostrom 2005; Ostrom et al. 1961). Second, it involves not only governmental actors, but also non-governmental actors who participate in the production and delivery of public goods and services. Third, policies and services have scales that go beyond political jurisdiction. Fourth, polycentric governance seeks cooperation and experimentation, promoting solutions that

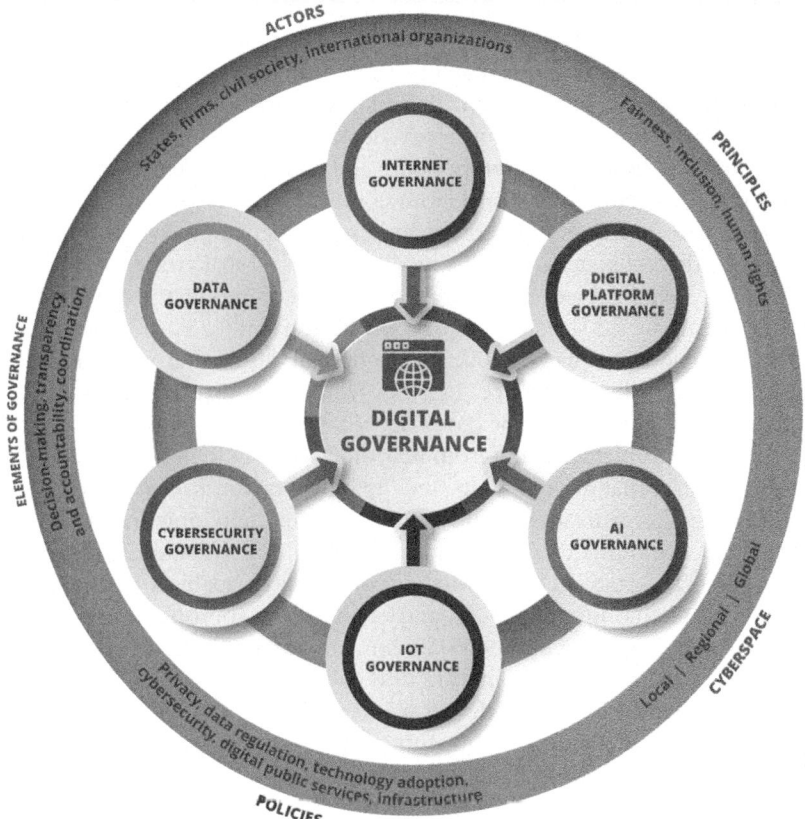

Fig. 4.1 The complex of digital governance institutions. (Source: Own elaboration)

solve the problems of coordination, conflict, and lack of accountability (Ostrom and Ostrom 1965; Oakerson 1999).

Polycentric governance operates in a context where institutions must be designed for multiple issues, with multiple autonomous and formally independent actors, choosing to act in ways that consider others, through processes of cooperation, competition, and conflict resolution (Ostrom 1991; Carlisle and Gruby 2017). This polycentric governance applied to the digital world focuses on forms of self-governing enterprise, achieving norms and rules that create another form of authority. An example of a social media self-governance institution is the global oversight board created by Facebook for content moderation. The company published a governing document that defines the board's mandate and describes its relationship to Facebook. It also defines the board's membership, governance, and decision-making authority (Facebook 2019).

Figure 4.1 shows the different issues and analytical instruments for the governance of the digital world. As a common-pool resource, the digital world requires polycentric governance instruments. As we showed in Chap. 2, the ecosystem of the digital world is global, with many local problems. Being subject to the tragedy of the commons, the governance of the digital world should be based neither on "more state" nor "more market." The dilemmas of collective action in the digital world require institutions of self-governance. In other words, the institutional grammar of digital governance must be based on collaboration, to provide solutions to the social problems arising from new digital technologies. This collaboration requires that governments, technology companies, civil society, and international organizations can establish norms and practices that are capable of designing a common grammar for different issues. Likewise, it is essential to establish a knowledge community capable of promoting interventions that minimize risks and establish the appropriateness of the application of new technologies in human organization, in its various aspects—political, social, economic, and cultural.

Thus, governance for the digital world exhibits a soft-policy characteristic, because it depends on elements of collaboration of diverse agents. Soft policy is a policy implemented in a cross-level coherence, requiring ratification and multilevel coordination (Torenvlied and Akkerman 2004). The coherence is a perspective that integrates the economic, social, and environmental and governance dimensions of policies in domestic and international policy-making scenarios. For example, cybersecurity policies demand coherence in procedures and standards, integrating social, economic, and political challenges to produce cyber-peace (Klimburg and Almeida 2019). Soft policy in the digital world demands ratification to produce outcomes. For example, regulatory policies based on international cooperation require ratification by states so that they can have international effects. Finally, policies for the digital world require multilevel coordination, involving global, national, and local actors to build solutions and evaluate results. For example, privacy policies require actors to coordinate in networked arrangements to ensure that they have a practical effect, as is the case with the GDPR in the European Union. The GDPR framework regulates the protection of personal data relating to individuals in the European Union with regard to the processing of personal data and on the free movement of such data.

The design of institutions to this end involves incorporating the governance elements applied to the digital world to promote new insights for institution building. In addition to having multiple decision-making actors, governance for digital must operate with non-territorial problems and in the context of global interdependence and overlapping. The overlapping is a concept to describe the jurisdiction or domain of decision makers (Carlisle and Gruby 2017).

In the digital world, the decision-making process passes through different layers of decisions taken by multiple actors. For example, the Internet is global with virtual connections that transcend national states. Decisions about Internet governance occur in layers applied to international, national, and local arenas. Social media exists in various societies, modifying communication. Thinking about soft-policy solutions means promoting the decentralization of authority and sharing problems and solutions to promote new practices and informal norms that govern the digital world and save the digital commons.

Governance for digital involves complex layers of decision making with multiple autonomous actors. A polycentric digital governance system comprises this holistic grammar, capable of understanding the actors, institutions, complexity of cyberspace, and the quest for cooperation and collaboration tools. The next section examines the elements of this polycentric digital governance system.

The Elements of Governance for Digital

In Chap. 2, we describe the elements of governance that arise in metagovernance analyses. Governance for the digital world requires that these elements gain specificity and concreteness to govern the digital world. An examination of the elements of digital governance indicates that the digital world's complexity requires complex solutions, open to experimentation and collaboration, to constitute a global grammar. This grammar should have a well-defined morphology and syntax, capable of incorporating various agents and institutions into one game that is nested in different issues, spaces, and problems. These elements aim for a holistic conception of digital governance, involving the various actors, spaces, and frameworks needed to govern the digital world in a polycentric format.

The institutional grammar of digital governance must consist of formal and informal rules that constrain the action of the actors at the same time as defining their stream. The grammar of digital governance requires a set of rules that favor collaborative decision-making structures, linked to instruments of coordination of the actors and rules that favor transparency and accountability. The linking of these elements makes it possible to constitute an institutional grammar capable of delimiting robust policies for the digital world.

The Decision-Making Process of the Digital World

The decision-making process is a central issue for digital governance. It incorporates the diverse governance spaces associated with the digital world and public governance, corporate governance, and international governance. Governance of the digital world incorporates the participation of governments, firms, international organizations, and civil society, so that decision making becomes a common element of Internet governance, cyberspace governance, governance of digital technologies, governance of social media, and data governance. Similarly, it incorporates problems that are global, regional, and local. That is, governance for the digital world is complex, comprised of multiple stakeholders, multiple forums, and multiple jurisdictions, building a polycentric governance pattern.

In addition, thinking about decision making in digital governance has two avenues. First, think about how autonomous systems make a decision and what the consequences of that decision are. Second, thinking about decision making means building decisions that result in rules and norms that regulate the digital world. In this section, we are interested in this second topic. Later, we will address the first topic when we discuss algorithmic transparency and accountability.

Governance for the digital world is complex, involving decisions about Internet infrastructure, cybersecurity issues, system and code standards, privacy, ICT use, regulation, and a host of other issues. These decisions are not made by a single authority (Leviathan) or an "invisible hand" of the market. It involves multiple stakeholders in different forums, requiring a multistakeholder approach to decision making and collaboration aspects taken globally, regionally, and locally. The decision-making process' complexity and the digital world's technocratic character preclude a broader participatory process. But the polycentric character delimits essential aspects for decisions to be taken and enforced robustly, to preserve the digital commons' shared character.

In the context of digital governance, multistakeholderism could be a characteristic of the polycentric model, because it strengthens the collaborative nature required by the digital governance process. This multistakeholder approach to digital governance involves "two or more classes of actors engaged in a common governance enterprise concerning issues they regard as public in nature and characterized by polyarchic authority relations constituted by procedural roles" (Raymond and DeNardis 2015, 573). The multistakeholder approach is contained in the polycentric perspective of digital governance, where the various actors build decisions on various issues in the digital world. The complexity of the multistakeholder approach lies in the diversity of institutional formats, where each actor can take the definition of multiple forums involving states, intergovernmental organizations, firms, and civil society (Raymond and DeNardis 2015).

Diverse institutional formats imply many decision centers, with multiple jurisdictions and a complex regime with overlapping subject matter. For governance of the digital world, Nye identifies 21 different organizations that incorporate states, firms, international organizations, and civil society. All these organizations participate in decisions about the digital world and are the authority to set rules and standards (Nye 2014). The digital governance regime's complexity creates several controversies about how the Internet should be controlled, as well as how technical standards, security formats and misinformation handling, and inclusion policies and institutional standards should be defined. Many of these organizations overlap and diversify their procedural authority.

The institutional grammar of digital governance works with the variation between different centers of decision and overlapping instances. Multiple constituted forums enable informed deliberation, so that they are competent to make technical decisions. Decision-making extends into diverse procedural arrangements of authority, making digital governance complex.

The decision-making process for the digital world has a collaborative element between the various actors. Beyond the multistakeholder perspective, digital governance requires elements of collaborative governance. Collaborative governance brings multiple stakeholders together in a common forum with public agencies to engage in consensus-oriented decision making (Ansell and Gash 2008). We argue that collaboration is a central element for polycentric governance of the digital world. It should consider the following:

> **ICANN—A Multistakeholder Perspective**
> The Internet Corporation for Assigned Names and Numbers (ICANN) is an organization responsible for Internet Protocol (IP) address space allocation, protocol identifier assignment, generic (gTLD) and country code (ccTLD) Top- Level Domain name system management, and root server system management.
> ICANN is a multistakeholder body that operates at the international level. The five Regional Internet Registries (RIRs) manage the distribution of number identifiers allocated by the Internet Assigned Numbers Authority (IANA). They are multistakeholder organizations that operate regionally.
> ICANN is a permanent organization with a multistakeholder structure for coordinating the Internet's naming system. In ICANN, all stakeholders (for example, governments, private companies, civil society, academia, and the technical community) get involved in the discussions. However, governments participate in ICANN processes only in an advisory capacity; they do not vote in the ICANN board's decisions. This is an example of a multistakeholder process under private sector leadership (Kleinwächter and Almeida 2015).
> The ICANN's coordination work is essential for the governance of the digital world, because it enables the construction of standards and protocols that guarantee machine-to-machine communication in cyberspace. The Internet fundamental structure of cyberspace operations depends on these international arenas and their collaboration.

- *the starting conditions of collaboration*, involving the resources of power, knowledge, and asymmetry between the actors, incentives for participation, and the trajectory of conflict and cooperation that define the initial levels of trust among stakeholders;
- *the collaborative process*, involving face-to-face dialogue between participants and negotiation, building trust between participants, commitment to process (with mutual recognition of interdependence), shared conceptions of the process and openness to multiple gains, and providing a shared understanding about rules and production of intermediate outcomes.

- *political leadership* that facilitates stakeholder involvement and participation; and
- *institutional design* that includes transparent rules, participatory inclusiveness, forum exclusiveness, and clear participant roles.

Polycentric structures increase transaction costs to decide, along with accountability failures, because it relies on multistakeholder participation and requires collaborative structures. Transaction costs for decision making increase, as polycentric governance becomes more geographically dispersed, between dispersed issues and overlapping perspectives. Similarly, there is no central authority—whether state, market, or society—to whom the various forums should be accountable (Carlisle and Gruby 2017). The polycentric character of governance for the digital world requires overlapping decision making to generate a complex combination of multiple levels and multiple types of organizations, creating multiple decision-making units that lower transaction costs and lack accountability.

Decision making in the digital world is more complex than conventional public policy management, because it is global and dispersed, and it requires multiple collaborative stakeholders to provide critical support for rules, norms, and principles of governance for the digital world. In the decision-making process, the grammar of digital governance does not have a single syntax, but a participatory and autonomous authority based on the collaboration and capacity with which organizations in this complex regime can support the existence of institutions.

As a governance element, the decision-making process in the digital world is complex and polycentric, with diverse institutional formats and perspectives. The capacity to take collective action in a participatory process is essential for solving the problems of the tragedy of the digital world, managing the risks of using digital tools, and defining the conditions for governance of the different elements of the digital world—the Internet, cyberspace, datafication, social media, and digital technologies.

Algorithmic Transparency and Accountability

Autonomous systems, using different types of algorithms, make decisions that affect society in many ways. Consequently, algorithmic transparency is an important requirement for the governance of the digital world. Algorithms make decisions without human intervention that directly affect society. Algorithmic decision systems perform predictions, ranking, and classifications and make decisions about people using various digital tools (Domingos 2015). These tools learn and refine their decisions according to large datasets and historical trends, reinforcing the epistemic character of computer systems. Algorithmic decision systems are used to predict and accumulate data and history to enhance decision-making skills.

By simulating a human decision system, machine learning can reinforce the role of past events. This may not lead to better decisions, because the use of past data and decision history has the potential to reproduce human and social biases. Despite the promise of neutrality of the autonomous decision-making systems, they are not neutral, when we consider that the combination of calculation, processing, and reasoning may reproduce decision systems designed by humans. Biases result in possible inequalities, such as spatial segregation, racism, sexism, or unequal distribution of resources (Eubanks 2018). As discussed in Chap. 3, algorithms can reinforce a technocratic character of human organizations and they are a central issue in the process of governance for the digital world (Janssen and Kuk 2016).

Algorithms are black boxes and challenge governance processes for the digital world. Black-boxed algorithms can limit opportunities, retract services, and produce technological redlining. This new type of inequality occurs when algorithms not only produce unequal results, but also replicate known inequalities, constituting a systematic means for exclusion of parts of society (Noble 2018). In addition, black-boxed algorithms may hamper society's trust, generating the reproduction of governance problems without digital technologies (Shull 2019).

The problem of algorithmic transparency brings a new challenge. In democracies, it is the citizen's right to know how politicians and bureaucrats make decisions affecting their lives (Filgueiras 2016). The right to know is a fundamental human right, and it should apply to algorithms, too (Pasquale 2015). Algorithmic transparency is the element from which citizens can know how autonomous decision systems make choices that affect their lives (Diakopoulos 2016; Hildebrandt 2012).

Algorithms embed an epistemic aspect, in which the system designers and development engineers set up their preferences to the machine. This makes algorithmic transparency a citizen's right to know how a specific decision has been achieved, turning algorithm transparency into a necessary condition for governing the digital world. For example, it is society's right to know how an algorithm of a government system distributes benefits to citizens. What criteria does this system use and how is the data processed so that society can measure the fairness of that governmental action?

The transparency of algorithms enables society to review autonomous decision systems that affect it, to be able to criticize and act to make systems more democratic. Different forms of transparency should be able to verify whether the expected results are based on the most basic principles of collective life. Algorithms, in their various modalities and possibilities, are complex and involve multiple layers of systems, encoding for thousands or billions of variables and thousands of data points. Algorithmic transparency has limits on the possibilities of secrecy—in both governments and firms—and on the citizen's cognitive capacities to create understanding (Crain 2018; Ananny and Crawford 2018).

In addition to the asymmetry problem between technology producers and users, governments and firms work with the possibility that secrecy may be justified in the context of a democratic order. At first glance, secrecy and publicity are antagonistic terms. However, some public policies require secrecy to be effective. Similarly, intellectual property requires firms to keep industrial secrets, including algorithms that make decisions about various forms of services—public or private—such as commerce and industrial technologies. Companies such as Google, Amazon, Facebook, and Microsoft keep their algorithms an industrial secret, not allowing any publicity for the decisions or recommendations they make to users.

On the other hand, society requires the fulfillment of the right to know. The algorithmic transparency demands a solution that allows secrecy to be constituent to some public policies. Thompson (1999) formulates the following principles concerning the issue of secrecy in democratic politics: (1) secrecy is only justified when promoting a democratic discussion on the merits of a public policy, and (2) secrecy is justified if the citizens and their representatives can deliberate upon it. The dilemma is solved by creating a first order of secrecy, which is then accompanied by a second order of publicity (Thompson 1999). That is, the decision to a secret must be made publicly.

In addition to the problem of secrecy, algorithmic transparency is hampered by citizens' cognitive limitations. Algorithms involve complex code, system layers, and technical language. Not all citizens are able to build an understanding of what that algorithm intends to do and how it came to that result. First, algorithms modify variables and data by their learning ability, making reverse engineering impossible and making various aspects of systems impossible to disclose (Diakopoulos 2016). Second, algorithmic transparency favors seeing over understanding (Ananny and Crawford 2018). Transparent algorithms do not necessarily promote understanding. Seeing the algorithm at work is not a sufficient condition for experiencing the various layers of systems. Understanding involves the possibility of experimentation and reflexivity, without which it turns into simple performance, without critical and applicable content to understand how algorithms work and behave.

The cognitive asymmetries between technology producers and users are high, making algorithmic transparency difficult. In addition to being very difficult to reverse engineer a code of complex systems, with multiple layers and functions, a citizen will hardly understand the commands and the logical chain of a simple coding. Algorithmic transparency is hampered by this asymmetry of knowledge. Many organizations have acted as knowledge brokers, in order to act as intermediaries between technology producers—meaning companies and governments, and users—meaning citizens (Xu et al. 2014). Organizations such as Data & Society, based in the United States, act as intermediaries between the production of technology and users, identifying biases, problems, and impacts of the adoption of technologies in everyday life. Likewise, several research centers in the technology area have acted as knowledge brokers, identifying biases in algorithms, and analyzing the results of autonomous systems and the impact of the use of digital technologies for society.

Algorithmic transparency is required to govern the digital world. However, it needs elements of accountability so that it can produce correct outcomes in the context of democracies. Accountability within political systems implies a governance process based on feedback, learning from experience and informed consent of the governed (Olsen 2017). The same reasoning can be applied to the digital world. Citizens and various users of digital tools are not the initial authors of algorithms or digital tools, even if they are based on co-creation formats. Nevertheless, citizens should not be viewed as powerless in this process. Accountability is not to be confused with transparency. Transparency is a governance tool that facilitates accountability. Transparency reduces the information gap between principals and agents and provides the conditions under which accountability can be performed in the context of an order and produce publicity (Filgueiras 2016).

Accountability involves the relationships between the forum and various actors who perform public actions. The forum can be an individual, organization, institution, or even an almost virtual entity such as public opinion, where the public policies and actions implemented by actors are discussed and judged. As a type of social relationship, actors have an obligation to justify their actions in the forum, which judges them from compliance with the rules and legitimacy of the action taken publicly. Accountability involves a complex process of justification and judgment. The relationship between actor and forum is one in which the actor is required to explain and justify his conduct. The forum can pose questions and pass judgment, and the actor may face consequences (Bovens 2007).

Digital public policies and services, as well as a variety of business services, have diverse consequences in public life. For example, social media has a strong influence on public communication (Gillespie 2010). Algorithms decide about human life with big data mechanisms. Cybersecurity problems affect all users of networks and digital services. All these issues make accountability crucial for regulating, coordinating, and controlling organizations (Christensen and Lægreid 2011; Papadopoulos 2010). Organizations are embedded in a dense web of accountability, where they must explain and justify their decisions and performance to various forums (Bovens et al. 2014).

Algorithm accountability seeks not only to disclose the algorithms, it also seeks to constitute responsibility and judgment for any actions that harm the public interest. Algorithms are steeped in normative questions, especially regarding compliance with the rules and the legitimacy of decisions made by autonomous systems, or even corporate decisions of technology companies related to data governance (Binns 2018). For example, Facebook CEO Mark Zuckerberg responded to a series of actions stemming from the use of private citizen data in the Cambridge Analytica scandal. The case refers to the 2016 presidential election process in the United States, which was answered by the American Congress. Likewise, the use of private data impacted the Brexit referendum, causing him to respond to the British Parliament. Parliaments functioned as principals, requiring the agent (Facebook) to be responsible for the acts performed. The institutions' response in the Cambridge Analytica case is configured as accountability limited to data governance problems. When applied to the dimension of the algorithms, the problem becomes more complex (Berghel 2018). In both cases, the principle is at a disadvantage of information and knowledge, making judgment difficult and creating governance failures.

The main limit about algorithmic accountability is the fact that the forum may not always be able to judge the agent's actions. As already stated for the problem of algorithmic transparency, algorithms are dynamic and comprised of several layers of systems based on a technical order. Algorithmic accountability is a matter of experts, reducing the forum's scope in the process of promoting the accountability of algorithms and its designers and developers. Accountability applied to algorithms is subject to controversy and scrutiny over the values and principles involved; it permanently requires justification of being used for the public good (Binns 2018).

The forum must be constituted with the authority to demand explanations from the actors, in a decentralized way and with possible measures for accountability (Schillemans and Smulders 2015). Accountability effectiveness requires institutions dedicated to this purpose, such as audits, inspections, the media, constituted political powers, and an autonomous judiciary. In addition, it is necessary to define a framework with which the web of accountability institutions can operate and for authors to promote compliance with the rules of the game. The goal is that the accountability system can generate learning and capacity of organizations to amend and resolve their own mistakes (Olsen 2017).

> **Algorithmic Justice**
> Algorithms can be applied to predict citizens' rights and benefits. Systems have been built to automate various government policies and agencies, in areas such as healthcare, criminal justice, and social work.
>
> In the area of criminal justice, algorithms have been employed for crime prediction and to estimate risk scores that are considered by judges during sentencing. These systems read lawsuits and produce classifications of individuals without society knowing how the algorithms work. In addition, they promote an excessive surveillance process that could punish the most vulnerable communities.
>
> Algorithm biases have become a key element for regulation, as they produce or reproduce social injustices. Insufficient databases, unethical behavior, or unintentional flaws in algorithm design produce most of the bias problems. Another problem that explains these biases and unfairly impacts algorithms is their instability. The behavior of most algorithms grows unstable, primarily because they were
>
> *(continued)*

(continued)
developed using data collected in a world before algorithms were used to make decisions. For example, changes in court-sentencing patterns could increasingly change the type and characteristics of offenders entering a prison. Those changes in the offender population eventually render any predictive algorithms increasingly irrelevant to the current population. Traditional risk-scoring systems or even human decisions face similar obstacles.

The definition of principles and practices for the design and coding of algorithms has been encouraged to reduce unfair biases and impacts. Technology companies should work to confirm algorithms' fairness and stability. This entails higher costs and has led to resistance from technology companies, which must review their procedures and teams, hire algorithm auditors, and ensure inclusion policies.

The problem with algorithmic accountability is that it is limited to the production of learning. Justifying and explaining the steps and how the algorithms perform the calculations lead to outcomes that make it difficult for society, and even for governance institutions—such as the executive, legislative, and judiciary branches—to appropriate systems and understand how they decide. Algorithmic accountability will be a technocratic matter, involving a technical language that is difficult for citizens and public managers to adopt and absorb. The algorithmic accountability framework should contain basic information about the algorithms and enable the experimentation of the codes and results achieved. Multiple centers in a polycentric governance perspective may significantly favor algorithmic accountability, in case conditions of transparency are given regarding the codes and procedures adopted in autonomous decision systems.

The framework discussed here points to different elements of algorithmic transparency that condition the possibility for the various forums—society and governance institutions—to demand justification and explanation from the actors responsible for the systems. It is important to highlight that this framework enables the necessary information, so that the forums can act to promote algorithmic accountability. This framework must be exhaustive enough to provide justification and explanation of the actors' action, and should contain the following elements:

- *Purpose and impact.* Citizens and institutions need to understand why the system exists, how it is used, by whom, where and when, and what kind of impact it will have on individuals and society.
- *Identification of the actors.* Indicate who the system owner is, and which organization is accountable, including the servers or employees in case of problems or need for information. Organizations should provide the information needed to identify responsible actors.
- *Oversight.* Systems should have independent third-party information that reviewed, tested, or audited the system before and during algorithmic decision-making operations.
- *Technical architecture.* Provide technical information about a broad context of the system and key connections with other systems, as well as procedures adopted for safety and security.
- *Input datasets.* Provide information on the procedures, steps for data collection, and the system variables used to produce outcomes.
- *Model and performance.* Provide information on how the system calculates, processes, and reasons about variables and how it does classification, prediction, and learning.
- *Output datasets.* Provide information on the results achieved by algorithmic decision systems and who, where, and when will have access and make use of the produced results.
- *Principles.* Provide information on principles that guide the design, and develop an operation of the system. This includes, for example, how the system addresses and deals with non-discrimination and inclusion of the various social actors, or how the system tests for bias and avoids redlining in the algorithmic decision-making process.
- *Explainability.* Provide answers that allow citizens to understand why the algorithmic system made a specific decision.
- *Human operator competences.* Produce insights into the involvement of human operators. In addition, explain the inclusion policies adopted and the various human perspectives adopted by these operators of digital tools.
- *Citizen rights.* Provide information on how citizens can claim their right to know.
- *Privacy and data protection.* Provide information about the privacy policy adopted, and how the system may affect privacy.

All elements of this framework aim to provide greater algorithmic transparency. To make these elements of transparency effective in the accountability process, it is necessary to organize various polycentric decision-making forums. They should be able to collect the required information and explanation in understanding various actors' actions regarding the construction of algorithmic decision systems. In addition, knowledge brokers can play a central role in producing evidence and its impact on society (Meyer 2010). Knowledge brokers in the digital world can play a central role, connecting evidence about the use and effects of digital technologies with the needs of the forum. In other words, knowledge brokers can provide the connection between the development of digital technology and the practical knowledge necessary for judgments by the principals to be issued. Knowledge brokers can facilitate this judgment by connecting the necessary knowledge so that governance decisions can be made and that eventual flaws or problems can be corrected, generating constant institutional learning.

The accountability process depends on how well the various polycentric decision-making forums are able to use public reason, so that digital governance institutions can justify their decisions with reference to a common viewpoint, including the principle of equal basic liberty, equal opportunities, and fair distribution of income and wealth. A principle of public reason for algorithmic accountability is essential for autonomous decision-making systems to produce legitimacy and justice (Binns 2018).

Coordination

The mantra that governance should promote stakeholder coordination to solve problems of collective action assumes that it improves the effectiveness of services and the various policies implemented. The discussion on coordination within governance arrangements starts from the discussion about policy networks (Rhodes 1997; Koppenjan and Klijn 2004) or the center of governments (Bouckaert et al. 2010). Policy networks need the coordination of various actors to produce sound decision and implementation processes. The center of government is the organizational arrangement that assigns political and institutional power to coordinate implementing agents and to constitute policy coherence. In either arrangement, the result is to produce effective policies. The assumption of a coordination theory is that there is an actor with agenda power who coordinates the other actors in the world of public organizations.

One of the major difficulties for coordinating actors in the digital world is that there is no central authority in the governance structure. The polycentric and decentralized character of digital governance requires coordination to take place through collaboration, replacing traditional forms of policy implementation. The challenge is that in digital governance, coordination is still needed. However, it happens without a central actor who takes on the task of coordinating the other actors. Coordination in the digital world is horizontal, depending on the interaction and relationships among the actors, with a high degree of horizontal coordination. In the digital world, horizontal coordination and network management are required to facilitate interactions between actors and create network structures.

A key feature of coordination in the digital world is the role of platforms to facilitate coordination. Collaboration takes place around shared values, dealing with networks and massive data production through platforms and the use of digital technologies. Coordination in the digital world occurs with the radical opening of the means of governance (Meijer et al. 2019). Digital governance implies the constitution of a collective and connected intelligence based on forms of self-organization (De Kerckhove 2001). The forms of coordination in the digital world are organized without leadership, with deconcentrated ways, and with distributed and connected interactions.

The polycentric nature of governance for the digital world implies innovative forms of coordination, based on using technology itself to solve collective action problems. The institutional design of governance for digital should facilitate the use of mechanisms for collaboration, both in the decision-making process and in coordinating solutions for collective action problems.

The governance elements applied to the digital world are essential to the ways in which technology is governed to make digital commons manageable, and with benefits to society. Coordination, decision making in the governance process, transparency, and accountability mechanisms are essential for promoting resilience in the digital world, constituting forms of self-governance and capacities to solve the different public problems created by the evolution of digital technologies.

Adaptive Capacity and Self-Governance

The central feature of the digital world is the availability of digital commons to society in cyberspace that can be explored or exploited by governments or firms. The impact of the digital world stems from a set of shared resources, which require norms—formal and informal—capable of promoting algorithmic transparency, accountability, decentralized and complex decision making in governance processes, and forms of coordination that are capable of establishing the effectiveness of the rules of the game.

Promoting the digital world's resilience entails governance mechanisms capable of facing permanent turbulence, so that its benefits can positively affect society. Resilience is the ability to adapt to adverse situations. The turbulence inherent in the context of the digital world, characterized by permanent change, requires constructing a dynamic resilience, underpinned by flexible institutional arrangements and organizations, capable of absorbing the digital world's complexity (Ansell and Trondal 2018) and contemporary challenges.

In addition to promoting resilience for institutions, governance for the digital world preserves digital commons, and thus requires digital development that incorporates social, economic, and political aspects. This development creates an innovative form of cooperation, especially in an increasingly shared and open dynamic. Governance for digital world arrangements will always be underpinned by dynamic equilibrium. Responding to changing conditions and technologies in the digital world will always require a new balance and reshaping of governance arrangements. Each remodel of governance arrangements must retain its basic functions and components, applying these to emerging technologies and turbulent contexts.

For equilibrium to be renegotiated, digital governance arrangements must be adaptive. Decisions will always occur within contexts of uncertainty, complexity, and substantial technological constraints, as well as conflicting human values and interests (Dietz et al. 2003). Adaptive capacities are defined as "the ability of a resource governance system to first alter processes and if required to convert structural elements as [a] response to experienced or expected changes in [the] societal or natural environment" (Pahl-Wostl 2009, 355).

Adaptive capabilities are enhanced when organizations are able to (1) provide information; (2) deal with conflict; (3) induce rule compliance; (4) provide infrastructure; and (5) be prepared for change (Dietz et al.

2003). Providing information is the ability of governance arrangements to create good and trustworthy information about various aspects of the digital world, especially the use of technology. Conflict is inherent to the digital world, because of different interests and values. However, it can be manageable if there are resources for conflict resolution within governance structures. Participating in open forums can be a great way to resolve conflicts and build rule adherence. Inducing compliance means that the rules—formal and informal—are followed by the stakeholders, considering the actors' expected behavioral patterns. Providing infrastructure is to put technology in favor of monitoring applied technologies, and enabling intervention when needed. Finally, being prepared for change is to create institutions that can adapt to more flexible norms for application in light of changing knowledge (Dietz et al. 2003).

Dietz, Ostrom, and Stern's reasoning applies to the tragedy of the commons regarding environmental issues. The same reasoning for the digital world ecosystem can be realized, as digital commons are managed on complex systems and require adaptive capacity to build resilience. The resilience of the digital world provided by polycentric governance makes it possible to build robust policies to address the different aspects of change. Changes in the digital world are caused by turbulence and technologies' evolution, which increase risks of exploitation. Adaptive capacities require robust policies that can handle and manage digital commons and mitigate existing risks. The robustness of a public policy for the digital world means that the actors can maintain the objectives and expected outcomes in a context of structural or procedural change (Ostrom 1990; Goodin 1996; Capano and Woo 2018).

Automating Poverty
The automation of public service implies the automation of all biases and problems of public administration. In developing countries, increasing automation of poverty has been observed.

Automated public services such as housing, food subsidies, and unemployment benefits are performed on digital platforms with AI layers that classify and predict poverty. Algorithms are used to suspend or cancel welfare payments to the poor. Or they require people

(continued)

(continued)

who do not have access to the Internet or digital skills to use applications so that they can access the benefit without any human intervention in the process.

Civil society organizations have denounced this process of suspending rights through opaque algorithms. Several cases have been reported in India, Australia, the United States, the United Kingdom, and other countries. In Brazil, for example, AI mechanisms have been used to suspend welfare benefits for extreme poverty without any explanation (Eubanks 2018).

The automation of poverty means widening inequalities and a scale of distrust with the use of digital tools in public policy, challenging the elements of digital governance.

The digital world's adaptive capacity will be as robust as its governance mechanisms are. As digital technologies advance, new opportunities are created, but new turbulent contexts emerge. Adaptive capacity is one in which, despite the advancement of technologies, institutions easily adapt to the new context, generating stability and robustness of policies. The various stakeholders should participate in the various decision centers, building deliberation and learning mechanisms. Mechanisms of transparency and algorithmic accountability should generate institutional learning, provide the ability to critique, and judge technologies to promote improvement. Coordination tools should work to manage conflicts, maintain the integrity of governance, and promote forms of collaboration and cooperation.

Resilience must be ensured by designing institutions in a complex, nonterritorial, multistakeholder format, and with the collaboration of bureaucratic hierarchies, the market, the state and civil society organizations, and technology users. The ability of the digital commons to deliver benefits to society depends on a polycentric and complex way of self-governing that is flexible and adaptive enough to bring technologies to society's benefit.

Governments have a responsibility to ensure the necessary regulatory frameworks for digital governance, as well as to ensure the proper use of Internet infrastructure. Governments must also build mechanisms to ensure cyber-peace and create instruments and standards for data governance and encourage technological policies that promote well-being. Technology companies must be accountable to the different governmental and international forums, as well as promoting transparent and appropriate solutions to the normative principles that guide digital governance. Civil society has the fundamental role of acting as a knowledge broker, connecting the knowledge produced by technology companies with the perspective of different users. Finally, international organizations must act in a global coordination effort for these actors, producing global consensus that can guide national policies. The resilience of digital governance institutions depends on this global effort involving these different actors in multistakeholder forums, for the implementation of policies to promote transparency, accountability, and horizontal coordination efforts.

Breaking the tragedy of the digital world means applying governance mechanisms within the Internet, cyberspace, digital technologies, social media, and the data and information collected and stored in the digital world. Maintaining the goal of freedom, no borders, and sharing the digital world will depend on how this world is governed, enabling more connections to knowledge.

References

Al-Badi, A., Tarhini, A., & Khan, A. I. (2019). Exploring Big Data Governance Frameworks. *Procedia Computer Science, 141*, 271–277. https://doi.org/10.1016/j.procs.2018.10.181.

Alexander, C. (1964). *Notes on the Synthesis of Form*. Cambridge: Harvard University Press.

Almeida, V., Doneda, D., & Monteiro, M. (2015). Governance Challenges for Internet of Things. *IEEE Internet Computing, 19*(4), 56–59. https://doi.org/10.1109/MIC.2015.86.

Ananny, M., & Crawford, K. (2018). Seeing Without Knowing: Limitations of the Transparency Ideal and Its Application to Algorithmic Accountability. *New Media & Society, 20*(3), 973–989. https://doi.org/10.1177/1461444816676645.

Ansell, C., & Gash, A. (2008). Collaborative Governance in Theory and Practice. *Journal of Public Administration, Research and Theory, 18*(4), 543–571. https://doi.org/10.1093/jopart/mum032.

Ansell, C., & Trondal, J. (2018). Governing Turbulence: An Organizational-Institutional Agenda. *Perspectives on Public Management and Governance, 1*(1), 43–57. https://doi.org/10.1093/ppmgov/gvx013.

Ansell, C., Trondal, J., & Øgård, M. (2017). *Governance in Turbulent Times.* Oxford: Oxford University Press.

Berghel, H. (2018). Malice Domestic: The Cambridge Analytica Dystopia. *Computer,* 51(5), 84–89. https://doi.org/10.1109/MC.2018.2381135.

Binns, R. (2018). Algorithmic Accountability and Public Reason. *Philosophy & Technology,* 31(4), 543–556. https://doi.org/10.1007/s13347-017-0263-5.

Bobrow, D. B., & Dryzek, J. (1987). *Policy Analysis by Design.* Pittsburgh: University of Pittsburgh Press.

Bolívar, M. P. R., & Meijer, A. J. (2015). Smart Governance: Using a Literature Review and Empirical Analysis to Build a Research Model. *Social Science and Computer Review,* 34(6), 673–692. https://doi.org/10.1177/0894439315611088.

Bouckaert, G., Peters, B. G., & Verhoest, K. (2010). *The Coordination of Public Sector Organizations.* London: Palgrave Macmillan.

Bovens, M. (2007). Analysing and Assessing Accountability: A Conceptual Framework. *European Law Journal,* 13(4), 447–468. https://doi.org/10.1111/j.1468-0386.2007.00378.x.

Bovens, M., Schillemans, T., & Goodin, R. E. (2014). Public Accountability. In M. Bovens, R. E. Goodin, & T. Schillemans (Eds.), *The Oxford Handbook of Public Accountability* (pp. 1–20). Oxford: Oxford University Press.

Capano, G., & Woo, J. J. (2018). Designing Policy Robustness: Outputs and Processes. *Policy & Society,* 37(3), 1–19. https://doi.org/10.1080/14494035.2018.1504494.

Carlisle, K., & Gruby, R. L. (2017). Polycentric Systems of Governance: A Theoretical Model for the Commons. *Policy Studies Journal,* 47(4), 927–952. https://doi.org/10.1111/psj.12212.

Christensen, T., & Lægreid, P. (2011). *The Ashgate Research Companion to New Public Management.* Farnham: Ashgate.

Crain, M. (2018). The Limits of Transparency: Data Brokers and Commodification. *New Media & Society,* 20(1), 88–104. https://doi.org/10.1177/1461444816657096.

Crawford, S., & Ostrom, E. (1995). A Grammar of Institutions. *American Political Science Review,* 89(3), 582–600. https://doi.org/10.2307/2082975.

Culkin, J. M. (1967). A Schoolman's Guide to Marshall McLuhan. *The Saturday Review,* 51–53, 70–72.

De Kerckhove, D. (2001). *The Architecture of Intelligence.* Berlin: Birkhäuser, Springer Science and Business Media.

DeNardis, L. (2014). *The Global War for Internet Governance.* New Haven: Yale University Press.

Diakopoulos, N. (2016). Accountability in Algorithmic Decision Making. *Communications of the ACM,* 59(2), 56–62. https://doi.org/10.1145/2844110.

Dietz, T., Ostrom, E., & Stern, P. C. (2003). The Struggle to Govern the Commons. *Science, 302*(5652), 1907–1912. https://doi.org/10.1126/science.1091015.

Domingos, P. (2015). *The Master Algorithm. How the Quest for Ultimate Machine Learning will Remake Our World.* New York: Basic Books.

Donahoe, E., & Metzger, M. M. (2019). Artificial Intelligence and Human Rights. *Journal of Democracy, 30*(2), 115–126. https://doi.org/10.1353/jod.2019.0029.

Eubanks, V. (2018). *Automating Inequality. How High-Tech Tools Profile, Police, and Punish the Poor.* New York: St. Martin's Press.

Facebook. (2019, September). *Oversight Board Charter.* Facebook. Retrieved from https://about.fb.com/wp-content/uploads/2019/09/oversight_board_charter.pdf.

Filgueiras, F. (2016). Transparency and Accountability: Principles and Rules for Construction of Publicity. *Journal of Public Affairs, 16*(2), 192–202. https://doi.org/10.1002/pa.1575.

Gasser, U., & Almeida, V. (2017). A Layered Model for AI Governance. *IEEE Internet Computing, 21*(6), 58–62. https://doi.org/10.1109/mic.2017.4180835.

GCSC. (2019). *Final Report of the Global Commission on the Stability of Cyberspace.* Retrieved from https://cyberstability.org/report/.

Gillespie, T. (2010). The Politics of 'Platforms'. *New Media & Society, 12*(3), 347–364. https://doi.org/10.1177/1461444809342738.

Goodin, R. E. (1996). Institutions and Their Design. In R. E. Goodin (Ed.), *The Theory of Institutional Design.* Cambridge: Cambridge University Press.

Helmond, A. (2015). The Platformization of the Web: Making Web Data Platform Ready. *Social Media + Society, 1*(2), 1–11. https://doi.org/10.1177/2056305115603080.

Hildebrandt, M. (2012). The Dawn of a Critical Transparency Right for the Profiling Era. In *Digital Enlightenment Yearbook 2012* (pp. 41–56). IOS Press. https://doi.org/10.3233/978-1-61499-057-4-41.

IEEE. (2019). Ethically Aligned Design. *The IEEE Global Initiative on Ethics of Autonomous and Intelligent Systems.* Institute of Electrical and Electronics Engineers (IEEE). Retrieved from https://standards.ieee.org/content/ieee-standards/en/industry-connections/ec/autonomous-systems.html.

Janssen, M., & Kuk, G. (2016). The Challenges and Limits of Big Data Algorithms in Technocratic Governance. *Government Information Quarterly, 33*(3), 371–377. https://doi.org/10.1016/j.giq.2016.08.011.

Kleinwächter, W., & Almeida, V. (2015). The Internet Governance Ecosystem and the Rainforest. *IEEE Internet Computing, 19*(2), 64–67. https://doi.org/10.1109/MIC.2015.49.

Klimburg, A., & Almeida, V. (2019). Cyber Peace and Cyber Stability: Taking the Norm Road to Stability. *IEEE Internet Computing, 23*(4), 61–66. https://doi.org/10.1109/MIC.2019.2926847.

Koppenjan, J., & Klijn, E. H. (2004). *Managing Uncertainty in Networks: A Network Approach to Problem Solving and Decision Making.* London: Routledge.

March, J., & Olsen, J. P. (1984). The New Institutionalism: Organizational Factors in Political Life. *American Political Science Review, 78*(3), 734–749. https://doi.org/10.2307/1961840.

Meijer, A. J., Lips, M., & Chen, K. (2019). Open Governance: A New Paradigm for Understanding Urban Governance in an Information Age. *Frontiers in Sustainable Cities, 1*(3), 1–9. https://doi.org/10.3389/frsc.2019.00003.

Meyer, M. (2010). The Rise of the Knowledge Broker. *Science Communication, 32*(1), 118–127. https://doi.org/10.1177/1075547009359797.

Millard-Ball, A. (2016). Pedestrians, Autonomous Vehicles, and Cities. *Journal of Planning Education and Research, 38*(1), 6–12. https://doi.org/10.1177/0739456X16675674.

Noble, S. U. (2018). *Algorithms of Oppression. How Search Engines Reinforce Racism.* New York: New York University Press.

Nye, J. S. (2014). *The Regime Complex for Managing Global Cyber Activities.* Global Commission on Internet Governance, Paper Series 1. Retrieved from https://www.cigionline.org/sites/default/files/gcig_paper_no1.pdf.

Oakerson, R. J. (1999). *Governing Local Public Economies: Creating the Civic Metropolis.* Ithaca, NY: ICS Press.

Olsen, J. P. (2017). *Democratic Accountability, Political Order, and Change. Exploring Accountability Processes in an Era of European Transformation.* Oxford: Oxford University Press.

Ostrom, E. (1990). *Governing the Commons: The Evolution of Institutions for Collective Action.* Cambridge: Cambridge University Press.

Ostrom, V. (1991). *The Meaning of American Federalism.* San Francisco, CA: Institute for Contemporary Studies Press.

Ostrom, E. (2005). *Understanding Institutional Diversity.* Princeton: Princeton University Press.

Ostrom, V., & Ostrom, E. (1965). A Behavioral Approach to the Study of Intergovernmental Relations. *Annals of the American Academy of Political and Social Science, 359*(1), 135–146.

Ostrom, V., Tiebout, C. M., & Warren, R. (1961). The Organization of Government in Metropolitan Areas: Theoretical Inquiry. *American Political Science Review,* 55(4), 831–842. https://doi.org/10.1017/S0003055400125973.

Pahl-Wostl, C. (2009). A Conceptual Framework for Analysing Adaptative Capacity and Multi-Level Learning Process in Resource Governance Regimes.

Global Environment Change, 29(3), 354–365. https://doi.org/10.1016/j.gloenvcha.2009.06.001.

Papadopoulos, Y. (2010). Accountability and Multi-Level Governance: More Accountability, Less Democracy? *West European Politics, 33*(5), 1030–1049. https://doi.org/10.1080/01402382.2010.486126.

Pasquale, F. (2015). *The Black Box Society: The Secrets Algorithms That Control Money and Information*. Cambridge: Harvard University Press.

Raymond, M., & DeNardis, L. (2015). Multistakeholderism: Anatomy of an Inchoate Global Institution. *International Theory, 7*(3), 572–616. https://doi.org/10.1017/S1752971915000081.

Rhodes, R. A. W. (1997). *Understanding Governance. Policy Networks, Governance, Reflexivity, and Accountability*. Buckingham: Open University Press.

Schillemans, T., & Smulders, R. (2015). Learning From Accountability?! Whether, What, and When. *Public Performance & Management Review, 39*(1), 248–271. https://doi.org/10.1080/15309576.2016.1071175.

Shull, A. (2019). Governing the Cyberspace During a Crisis in Trust. In Centre for International Governance Innovation (Ed.), *Governing Cyberspace during a Crisis in Trust. An Essay Series on the Economic Potential—And Vulnerability—Of Transformative Technologies and Cyber Security* (pp. 4–8). Waterloo: Centre for International Governance Innovation.

Siddiki, S., Heikkila, T., Weible, C. M., Pacheco-Vega, R., Carter, D., Curley, C., Deslatte, A., & Bennett, A. (2019). Institutional Analysis with the Institutional Grammar. *Policy Studies Journal*. Early View. https://doi.org/10.1111/psj.12361.

Thompson, D. (1999). Democratic Secrecy. *Political Science Quarterly, 114*(2), 181–193. https://doi.org/10.2307/2657736.

Torenvlied, R., & Akkerman, A. (2004). Theory of 'Soft' Policy Implementation in Multilevel Systems with an Application to Social Partnership in the Netherlands. *Acta Politica, 39*(1), 31–58. https://doi.org/10.1057/palgrave.ap.5500046.

Xu, Z., Ramanathan, J., & Ramnath, R. (2014). Identifying Knowledge Brokers and Their Role in Enterprise Research through Social Media. *Computer, 3*(1), 26–31. https://doi.org/10.1109/MC.2014.61.

CHAPTER 5

Conclusions

Abstract This chapter summarizes the main contributions of this work. The framework presented in this book is a step toward a broader understanding of governance challenges for the digital world. Theories borrowed from different fields of science contribute with knowledge to construct digital governance frameworks.

Keywords Governance • Tragedy of commons • Institutional design • Common-pool resources • Multistakeholder • Coordination • Transparency • Accountability

The digital world comprises technologies that connect people, information, services, and things. Theories borrowed from different areas of science—such as economics, political science, and environmental science—contribute with blocks of knowledge that can be harnessed to construct governance frameworks for the digital world. Among the building blocks, we can cite the metaphor of the tragedy of the commons, concepts of polycentric governance, and the multistakeholder model.

The tragedy of the digital world undermines society's confidence in institutions, creating uncertainties and a context of permanent turbulence. These turbulent junctures compromise how well human organization functions, requiring new solutions to govern a world in which its organizations are ever-changing. The use of digital technologies can benefit society as a whole by creating new solutions to central public problems and promoting well-being. The potential of technologies to optimize services and public policy is enormous. However, exploitation of the digital world for private benefit compromises the way governments and private sector organizations can incorporate digital technologies into their procedures, policies, and services.

Governing the digital world challenges knowledge about government, polity, and policy. In this book, we offer an initial framework for institutional and policy design for the digital world. We conceptualize digital governance as the arrangement of the polycentric institutions in the digital world to govern in a legitimate, inclusive, and secure manner the use of digital resources to produce sustainable services and public policies implemented by governments and firms in a non-territorial and results-based manner.

The governance of the digital world requires a design that transcends the state and market dichotomy. The digital world cannot be governed only in the dynamics of centralization of power in state authority. The danger of this process is to create the colonization of the digital world, especially with state control over data and information. Total state control over data and information amplifies existing forms of surveillance. This colonization process also creates a new order of conflicts between states, so that the governance of the digital world becomes a central geopolitical problem (Couldry and Mejias 2019). The increasing nationalization of the Internet has been a trend in international forums and compromises the global character and free circulation of information, according to its original conception.

On the other hand, the digital world cannot be governed by market dynamics. Enabling technology companies to govern the digital world reinforces market failures, promotes the growing oligopolization of the digital world, and leaves data and information on a global scale under control of these companies. In fact, technological innovations are of concern to governments because there is a tendency toward oligopolization. Perspectives of industrial secrecy and the lack of an authority to impose rules makes societies subject to all sorts of disinformation and misinformation.

The theoretical corollary of the tragedy of the digital world poses the following problem, which we highlighted at the beginning of this book: How can society create a governance process for the digital world? Evidently, it is a complex question and any answer to it will be incomplete.

However, we can think of principles that organize basic ideas, bring the constitutive elements to this governance process, and guide political action. This political action can be aimed at building an ecosystem of the digital world that maintains its liberal functions, creates regulation and adequate rules to prevent its growing pollution, prevents the common-pool resources from being appropriated in an improper way, and makes shared resources (data and information) benefit the whole society, in a fair, correct, and oriented way to solve various social problems. Likewise, we need to think of institutions that are resilient enough to withstand all the turbulence caused by technological innovation. This is certainly not an easy task, and it is now at the center of the global political agenda (United Nations 2019).

The governance model for the digital world is neither "more state" nor "more market." Following the clues posed by the challenges of governing the commons, the model for digital governance is self-governance on a global scale, creating collaborative ways that involve multiple stakeholders in different forums, create mechanisms of effective transparency and accountability, and act in a coordinated way to mitigate the unintended effects of accelerated change promoted by technological innovations. A self-governing model must promote ethical principles and effective regulatory frameworks to govern the digital world and design institutions that are resilient to face all challenges.

Institutional design for the digital world refers to the capacity of governments, firms, civil society organizations, and international organizations to collaborate and jointly build solutions that enhance the transparency and accountability of algorithms, encourage coordinated efforts, and promote open and inclusive decision making. Policy design for the digital world must ensure that governance of the digital world is based and grounded firmly in democratic values and human rights, fostering the co-creation of services, cybersecurity as a common issue, and a process of regulation that protects the digital realm from world exploitation and makes it possible to govern policy for the benefit of society.

Preserving digital common-pool resources means constructing policies that guide policy makers toward the resilience of the human organization in its many aspects. Governments need to adapt to organizational changes caused by the digitization of society, considering aspects of the logic of appropriateness and not just the logic of cost-benefit; otherwise, it will maximize the risks that amplify turbulent situations.

Technological solutions such as AI, IoT, machine learning algorithms, platformization of government and business, datafication, and cybersecurity demand governance processes that drive digital transformation toward the preservation of values of liberal democracy and human rights. The analysis of the digital governance process should consider its fragmentation into different issues and institutional overlapping. The need for a holistic perspective on digital governance stems from the construction of an institutional design that considers multistakeholder issues and global problems.

Promoting multistakeholder coordination, transparency, and accountability, along with global collaboration and resilience to the emergence of new technologies provides a possibility to define a common institutional grammar for the digital world. The assumption that the digital world could be governed relies on the ability of actors to preserve digital commons and make the Internet, data, cyberspace, and platforms share knowledge and promote well-being. This means that the digital world must be more transparent and accountable, more coordinated, and use decision-making processes that ensure societal participation and deliberative perspectives to address insecurity and infrastructure issues, disinformation, discrimination, polarization, and diverse issues that arise and challenge the digital world.

This book will not end on these pages. The challenges of digital governance move forward as governments, companies, and individuals absorb and increase the use of new digital technologies. The way advanced technologies rapidly scale and change human organization makes a holistic view of digital governance necessary to uphold the values of democracy and human rights. Here we offer the essentials for designing institutions and policies that foster and encourage society's ability to govern the digital world. The framework presented in this book is a step toward a broader understanding of governance challenges for the digital world. Understanding actors, values, policies, and spaces is critical for designing solid institutions that achieve the values of democracy and human rights in the digital world.

References

Couldry, N., & Mejias, U. (2019). Data Colonialism: Rethinking Big Data's Relation to the Contemporary Subject. *Television & New Media*, *20*(4), 336–349. https://doi.org/10.1177/1527476418796632.

United Nations. (2019). *The Age of Digital Interdependency*. Report of the Secretary-General's High-level Panel on Digital Cooperation. New York: United Nations. Retrieved from https://www.un.org/en/pdfs/DigitalCooperation-report-for%20web.pdf.

Index

A
Accountability, 5, 8, 22, 23, 25, 29, 30, 32–34, 47, 53, 54, 75, 76, 81, 84, 87–93, 95–97, 99, 107, 108
Agencification, 49, 55
Agranoff, R., 58
Ahn, M., 53
Airbnb, 2, 45
Akkerman, A., 82
Al-Badi, A., 79
Alexander, C., 77
Algorithms
  algorithmic accountability, 91–93, 99
  algorithmic transparency, 84, 87–95
  automation, 98, 99
  bias, 2, 48
  injustice, 11
Alkhatib, A., 47, 48
Allcott, H., 55
Almeida, V., 32, 46, 52, 79, 82
Amazon, 2, 11, 17, 19, 89
Ananny, M., 89, 90
Andrews, L., 48
Ansell, C., 59, 61, 78, 85, 97

Apple, 2, 11, 17
Appropriateness, 60, 82
Artificial intelligence (AI), 2, 9, 11, 13, 15, 19, 34, 44, 51, 54, 63, 79, 98
  machine learning, 45, 52, 108
Attard, J., 53

B
Balutis, A. P., 50
Bannister, F., 21, 50
Barlow, J. P., 8, 9, 16
Bartje, J., 52
Beck, U., 62
Bell, S., 24
Benkler, Y., 10, 13, 15
Berghaus, S., 57
Berghel, H., 91
Berman, F., 35
Bernstein, M., 47, 48
Bertot, J., 51
Bindu, N., 18
Binns, R., 91, 92, 95
Blockchain, 9, 12, 53

Bobrow, D.B., 77
Bolívar, M.P.R., 76
Bouckaert, G., 25, 95
Boulos, K., 46
Bovaird, T., 26
Bovens, M., 91
Bowers, J., 13
Boyd, D., 21
Brass, I., 53
Brazil, 31, 47, 51, 97
Bretschneider, S., 53
Brewer, G.A., 55
Budish, R., 30

C

Cambridge Analytica, 13, 91
Capacities, 18, 22, 23, 25, 29–33, 35, 44, 46, 50, 52, 53, 56, 58, 61, 62, 64, 87, 89, 92, 96–99, 107
Capano, G., 98
Caplan, R., 21
Carlisle, K., 81, 83, 87
Carlos-Roca, L.R., 34
Cellphone tracking, 20
Centeno, M., 31
Cerf, V.G., 35
Chen, D., 59
Chen, Y.C., 58
Chen, Y.F., 19, 27, 46
China, 20, 45–47
Christensen, T., 91
Chun, S.A., 54
Ciborra, C., 60
Cingolani, L., 30
Civil society, viii, 3, 13, 28, 30, 35, 57, 80, 82, 84–86, 99, 100, 107
Clark, D., 10
Co-creation, 26, 90, 107
Cohen, M.D., 57
Commons
  collective action, 15, 16, 18
  commodification, 3, 16, 18
  common-pool resources, 14, 15, 17–20, 82, 107
  digital commons, 13, 15–22, 27–29, 83, 85, 96–99, 107, 108
  exploitation, 15–17, 107
  public goods, 13, 14, 20
  tragedy of the commons, 2, 13–21, 82, 98, 105
Companies, 3, 4, 12, 13, 17, 20, 23, 28–30, 33, 51, 62, 63, 65, 76, 82, 89–91, 93, 100, 106, 108
Connolly, R., 21, 50
Coordination, 5, 24, 47, 54–56, 58, 81, 82, 84, 95–97, 99, 108
Couldry, N., 106
Crain, M., 89
Crawford, K., 89, 90
Crawford, S., 22, 77
Creemers, R., 47
Cukier, K., 12
Culkin, J. M., 75
Cyberspace
  cyberattacks, 15, 63
  cyber-peace, 82, 100
Cyert, R., 31

D

Dahlberg, L., 55
Dalton, R., 64
Danaher, J., 46, 48
Data
  big data, 44, 54, 91
  datafication, 12, 79, 87, 108
  data protection, 59, 94
Data & Society, 90
Dawes, S.S., 61
Decision making, viii, 1, 2, 8, 11, 12, 23–25, 29, 30, 32, 34, 35, 44, 45, 53, 55, 60, 63, 80, 82–88, 94–97, 107, 108
De Kerckhove, D., 96
Deloitte, 63, 64

Democracy
  elections, 13
  participation, 13, 55
  polarization, 17
DeNardis, L., 21, 79, 85
Design
  design institutions, 80, 107, 108
  policy design, 49, 106, 107
Diakopoulos, N., 11, 88, 89
Dietz, T., 97
Digital divide
  exclusion, 58, 61
  inequalities, 18
Digital governance
  AI governance, 34
  data governance, 13, 79, 84, 91, 100
  digital transformation, 45, 56, 59, 61, 108
  internet governance, 21, 79, 83, 84
  IoT governance, 34, 79
  platform governance, 79
  smart governance, 76
DiMaggio, P., 24
Disinformation, 12, 14, 15, 19, 31, 33, 106
Disruption, 9, 18, 20, 57, 60, 78
Domingos, P., 44, 47, 87
Donahoe, E., 32, 78
Donahue, J.D., 26
Douglas, M., 62
Dryzek, J., 77
Duff, A.S., 20
Duguit, L., 48
Dunleavy, P., 46, 55, 58, 61
Dye, T., 48

E
Earley, S., 58
Ecosystem, 13–15, 20, 33, 47, 55, 82, 98, 107
Effectiveness, 15, 23, 25, 50, 56, 57, 61, 64, 78, 92, 95, 97

Eggers, W.D., 52
Electronic Frontier Foundation (EFF), 78
Etzioni, A., 54
Eubanks, V., 87, 97

F
Facebook, 2, 10, 11, 17, 19, 20, 31, 45, 75, 82, 89, 91
Facial recognition, 34, 79
Ferro, E., 58
Filgueiras, F., 25, 50, 88, 90
5G, 21
Floridi, L., 3, 27
Fountain, J.E., 50
Frenken, K., 18
Fung, A., 25

G
Gash, A., 86
Gasser, U., 21, 52, 79
GCSC, 79
GDPR, 82
Gentzkow, M., 55
Geraghty, E., 46
Gil-Garcia, J.R., 56
Gillespie, T., 17, 21, 53, 79, 91
Gillingham, P., 46
Goodin, R.E., 77, 78, 97
Google, 2, 11, 17, 28, 45, 75, 89
Governance
  collaborative governance, 86
  corporate governance, 23, 26, 84
  global governance, 26, 30, 32
  interactive governance, 23
  meta-governance, 24–26, 83
  polycentric governance, 80–82, 85, 87, 93, 98, 105
  public governance, 26, 57, 84
  self-governance, 82, 96–100, 107

Greengard, S, 46
Group of Twenty (G20), 78
Gruby, R.L., 81, 83, 87

H
Hadfield, G., 17
Hanna, M., 13
Hansen, A.M., 61
Hardin, G., 2, 13, 18
Hardin, R., 64
Hate speech, 12, 17
Heald, D., 25, 54
Helmond, A., 52, 79
Hemmati, M., 30
Hess, C., 15–18
Hildebrandt, M., 88
Hood, C., 50
Howard, P.N., 21, 34
Howlett, M., 48, 49
Hsieh, T.C., 46
Human rights, vii, 2, 32, 34, 35, 45, 76, 78, 88, 107, 108

I
IBM Watson, 11
IEEE, 76, 78
Inclusion, 25, 29, 58–59, 85, 93
Inequalities, 7, 9, 16, 18, 88, 99
Inglehart, R., 64
Instagram, 10, 19, 31
Institutions
  capacities, 30–31, 50, 107
  institutional design, 76, 77, 80, 87, 96, 107, 108
  institutional grammar, vii, 21, 77–80, 82, 84, 85, 108
  norms, 35, 78
  rules, 25, 26
  standards, 4, 85
Internet, 1, 3, 4, 9–11, 13–17, 19–21, 29, 52, 54, 58, 79, 83–87, 99, 100, 106, 108

Internet Corporation for Assigned Names and Numbers (ICANN), 4, 30, 78, 86
Internet of things (IoT), 9, 12, 13, 15, 34, 35, 52, 79, 108
Isaak, J., 13

J
Janssen, D., 55
Janssen, M., 48, 54, 60, 87
Jarrahi, M.H., 18
Jessop, B., 24, 30
Jordan, T., 55
Justice, 11, 22, 25, 45, 46, 91, 94
Just. N., 46

K
Kang, J.S., 52
Kaplan, A., 10
Kennett, P., 22, 26
Kies, R., 55
Kitchin, R., 48
Kleinwächter, W., 30
Klijn, E.H., 23, 95
Klimburg, A., 79, 82
Knowledge, 2, 3, 9, 13–17, 25, 30, 33, 46, 59, 62, 63, 82, 86, 91, 98, 105, 106, 108
  knowledge brokers, 90, 95, 100
König, P.D., 47
Koppenjan, J., 23, 95
Kosters, M., 48
Kuk, G., 48, 87

L
Lægreid, P., 91
Latzer, M., 46
Layers, 3, 10, 47, 51, 83, 89, 90, 92, 98
Lee, L.F., 18
Legner, C., 57

Levi, M., 64
Linders, D., 58
Lindstedt, C, 54
Lipsky, M., 47
Luna-Reyes, L.F., 55, 56

M

Machine learning, 44, 45, 52, 88, 108
Madakam, S., 52
Maedche, A., 58
Mansell, R., 17
March, J.G., 4, 8, 23, 31, 60, 77
Margetts, H., 50, 53, 58, 61
Markets, 2, 21, 24, 26, 28, 32, 79, 82, 84, 87, 99, 106, 107
Masso, A., 61
Matthews, F., 29
Mayer-Schönberger, V., 12
McDermott, P., 53
McGuire, M., 58
Meijer, A.J., 18, 21, 58, 76, 96
Mejias, U., 106
Mergel, I., 57
Metzger, M.M., 32, 78
Meyer, M., 93
Microsoft, 2, 17, 89
Mikhaylov, S.J., 54
Milakovich, M.E., 48
Millard-Ball, A., 80
Mills, S., 18
Mondal, M., 17
Monitoring, 12, 15, 45, 54, 98
Moynihan, D. P., 49
Mueleman, L., 24
Myers, S., 53
Myers West, S., 12

N

Nagle, F., 13
Napoli, P.M., 46
Naurin, D., 54
NetMundial, 30

Networks, 3, 10, 15, 23, 24, 26, 28, 30, 31, 46, 52, 54, 59, 61, 82, 91, 95
Newton, K., 64
Noble, S.U., 88
Nord, J.H., 52
Nye, J.S., 26, 78, 79, 85

O

Oakerson, R.J., 81
Oakley, K., 27
Offe, C., 64
Olsen, J.P., 4, 8, 23, 25, 31, 56, 59, 60, 77, 90, 92
O'Neil, C., 11
Open government, 13, 53, 54
O'Reilly, T., 53
Organization for Economic Co-operation and Development (OECD), 4, 23, 30, 59, 78
Organizations, 3, 9, 43, 76, 105
  changing, 56–62
Osborne, S., 26
Ostrom, E., 4, 14–18, 20–22, 26, 27, 77, 78, 80, 98
Ostrom, V., 80, 81

P

Pagano, M., 26
Pahl-Wostl, C., 97
Palvia, S., 27
Papadopoulos, Y., 91
Park, A., 24
Pasquale, F., 88
Perl, A., 55
Perlroth, N., 63
Peters, B. G., 4, 5, 22–26, 50
Pierre, J., 4, 5, 23, 24
Platforms, 1–3, 9–11, 13, 15, 19, 20, 28, 31, 45, 53–54, 57, 76, 96, 98
  platformization, 51, 52, 79, 108
Policy

policy design, 49, 106, 107
policy tools, 43, 49, 50
soft policy, 82, 83
Powell, W., 24
Power, vii, 2–4, 8, 9, 11, 13, 17, 21, 31, 33, 45, 47–49, 61, 86, 92, 95, 106
Privacy, 11, 13, 21, 32, 33, 46, 54, 59, 79, 82, 84, 94
Public services, 2, 13, 24–26, 33, 34, 46–49, 51, 53–55, 58, 61, 64, 98

Q
Qiang, X., 20

R
Rahman, S., 17
Raymond, M., 85
Regulation, 4, 9, 21, 28, 30, 49, 50, 52, 60, 79, 84, 107
Resilience, 61, 65, 78, 80, 96–100, 107, 108
Rhodes, R.A.W., 23, 95
Risk, viii, 1, 2, 5, 21, 32–34, 44, 45, 61–65, 76, 80, 82, 87, 98, 107
risk assessment, 2, 34
Rosenau, J., 23, 26

S
Sahel, J.J., 30
Salamon, L., 23, 49, 50
Samuel, A., 44
Sandoval, R., 57
Sandoval-Almazán, R., 57
Satariano, A., 63
Scheerder, A., 59
Schillemans, T., 92
Schor, J., 18

Seo, D., 58
Servick, K., 45
Shackelford, S., 12, 53
Sharing
  sharing economy, 18
  sharing politics, 18
  sharing society, 18
Sharma, S., 27
Shull, A., 88
Siddiki, S., 77
Silicon Valley, 20
Simon, H. A., 31, 45
Smulders, R., 92
Social media, 3, 9–13, 15, 17, 19, 20, 28, 31, 52, 55, 63, 79, 82–84, 87, 91, 100
Sörensen, E., 23, 24
State, 2, 5, 8, 9, 13, 14, 20, 21, 24–28, 30, 32, 34, 44, 45, 49, 50, 53, 55, 56, 59, 60, 62, 64, 65, 79, 82, 83, 85, 87, 99, 106, 107
Stern, P.C., 98
Stoker, G., 22, 23, 26, 56
Stone, D., 23
Sunstein, C., 12, 17, 29
Surveillance, 12, 17, 20, 21, 33, 45, 46, 106
Sutherland, W., 18

T
Tammpuu, P., 61
Tang, G., 18
Tassabehji, R., 51
Thompson, D., 89
Torenvlied, R., 82
Torfing, J., 23, 24
Transparency, 5, 22, 23, 25, 30, 33–35, 47, 53, 54, 76, 84, 87–97, 99, 100, 107, 108

Trondal, J., 59, 61, 97
Turbulence, 9, 56, 59–61, 63, 76, 78, 98, 99, 105, 107
Twitter, 2, 10, 11, 17, 31, 45, 75

U
Uber, 2, 45, 75
United Nations (UN), 4, 7, 14, 30, 107
United States, 9, 20, 90, 91, 99
UN-ITU, 30, 78
UNPAN, 27

V
Values, 15, 22, 24–26, 29, 32, 53, 56, 57, 63, 64, 76–78, 92, 96, 97, 107, 108
Van der Heijden, J., 48
Van Dijk, J.A.G.M., 21
Vardi, M., 20
Veale, M., 53
Vial, G., 57
Volpin, P.F, 26

W
Waite, C., 18
Warren, M., 25

Weber, R.H., 52
WeChat, 20, 46, 75
Welchman, L., 28
Welzel, C., 64
Werbach, K., 10
Westerman, G., 56
WhatsApp, 10, 31, 45, 75
Wildavsky, A. B., 61, 62
Williamson, B., 46, 58
Woo, J. J., 98
World Bank, 23
World Health Organization, 14
Wu, D., 59
Wu, S., 59

X
Xu, Z., 89
Xue, M., 28

Y
YouTube, 2, 10, 11, 45, 75

Z
Zhao, H., 59
Zittrain, J., 13
Zuboff, S., 11, 17
Zuckerberg, Mark, 20, 91

The manufacturer's authorised representative in the EU is Springer Nature Customer Service Centre GmbH, Europaplatz 3, 69115 Heidelberg, Germany. If you have any concerns regarding our products, please contact ProductSafety@springernature.com

Printed and bound by CPI Group (UK) Ltd, Croydon, CR0 4YY

25/03/2026

02078205-0005